Nano Technology

Dr.P. Parthasaradhy

Dr.K.S. Chandragupta Mauryan

G. Gnaneshwar Kumar

Published by

ISBN 978-93-92537-93-6

Authors

Dr.P. Parthasaradhy | Dr.K.S. Chandragupta Mauryan | G. Gnaneshwar Kumar

Bonfring

309, 5th Street Extension, Gandhipuram,

Coimbatore - 641 012,

Tamil Nadu, India.

E-mail: info@bonfring.org

Website: www.bonfring.org

Preface

Welcome to the captivating world of nanotechnology! This book unfolds a comprehensive journey through the fundamentals of nano materials and technologies, offering a profound understanding of the intricacies and applications that shape this revolutionary field.

In the ever-evolving landscape of scientific revolutions, nanotechnology stands at the forefront, pushing the boundaries of what we thought possible. The periodic table becomes a playground for innovation, and the atomic and molecular realms unveil their secrets, setting the stage for the exploration of nano dimensions.

Our exploration begins with a foundational understanding of the atomic structure, molecules, and phases, navigating through the vast expanse of energy and dimensional space in Unit I. The top-down and bottom-up approaches serve as our guiding principles, leading us towards the heart of molecular nanotechnology.

Unit II delves into the world of molecular nanotechnology, exploring atoms through inference and harnessing the power of advanced tools like electron microscopes and scanning probe microscopes. The art of self-assembly takes center stage, showcasing the elegance of nature-inspired design.

Nano powders and nanomaterials take center stage in Unit III, where we unravel the secrets of their preparation using techniques such as plasma arcing, chemical vapor deposition, and sol-gels. From ball milling to the utilization of natural nanoparticles, we witness the versatility and applications of nanomaterials.

Unit IV opens the gateway to nanoelectronics, exploring the fabrication of integrated circuits, MEMS, NEMS, and the wonders of nano circuits. Quantum wires, quantum wells, and the fascinating DNA-directed assembly pave the way for revolutionary applications in electronics.

Finally, Unit V brings us to the forefront of applications, from MEMS and NEMS to coatings, optoelectronic devices, environmental solutions, and the groundbreaking field of nanomedicine.

Embark on this educational journey with us, where each page unravels the mysteries of nano materials and technologies. Whether you are a student, researcher, or enthusiast, this book aims to inspire and equip you with the knowledge to navigate the frontiers of nanotechnology.

Acknowledgement

In bringing this book to fruition, we are profoundly grateful to the countless individuals whose contributions have shaped its pages. The journey from conception to completion has been a collaborative effort, and we extend our heartfelt thanks to those who have played a pivotal role.

First and foremost, we express our deepest gratitude to the educators and researchers who have dedicated their expertise and knowledge to the field of nanotechnology. Your passion and commitment have inspired not only the content of this book but also the minds of future innovators.

We extend our appreciation to the students and readers who embark on this educational voyage. Your curiosity and enthusiasm fuel the essence of this work, and it is our hope that this book serves as a valuable resource in your exploration of nano materials and technologies.

A special acknowledgment is reserved for the institutions and organizations that have supported and facilitated the research and compilation of this material. Your commitment to advancing scientific knowledge is invaluable, and we are honoured to be part of the collective effort.

We are indebted to the pioneers in the field of nanotechnology whose groundbreaking research laid the foundation for the content presented here. Your trailblazing spirit has illuminated the path for future generations of scientists and engineers.

Lastly, we express our gratitude to friends and family members for their unwavering support throughout this endeavor. Their encouragement has been a source of strength, and we are fortunate to have such a dedicated network.

To all those who have contributed, directly or indirectly, to the realization of this book, we extend our sincere thanks. May this work inspire and contribute to the ongoing dialogue in the dynamic realm of nanotechnology.

TABLE OF CONTENTS

Unit - I

Background of Nanotechnology

1.1. Introduction

Nanotechnology is science, engineering, and technology conducted at the nanoscale, which is about 1 to 100 nano meters.

Nanotechnology is the understanding and control of matter at dimensions between approximately 1 and 100 nano meters.

Nanotechnology is the manipulation of matter on the nanoscale.

The kinds of product that could be built will range from microscopic, very powerful computers to super strong materials ten times as strong as steel, but much lighter too, food to other biological tissues. All these products would be very inexpensive because the molecular machines that built them will basically take atoms from garbage or dirt, and energy from sunshine, and rearrange those atoms into useful products, just like trees and crops take dirt, water and sunshine and rearrange the atoms into wood and food.

- Nano is a Greek word that means **'dwarf'**
- The word 'nano' is used to refer to 10^{-9} *size of atom is* 10^{-10}
- At the nanoscale, strange things happen to materials: – their properties can change.
- Carbon in the form of graphite (i.e., pencil lead) is soft, at the nano-scale, can be stronger than steel and is six times lighter.

Ex: Nano-scale copper is highly elastic metal at room temperature, stretching to 50 times its original length without break.

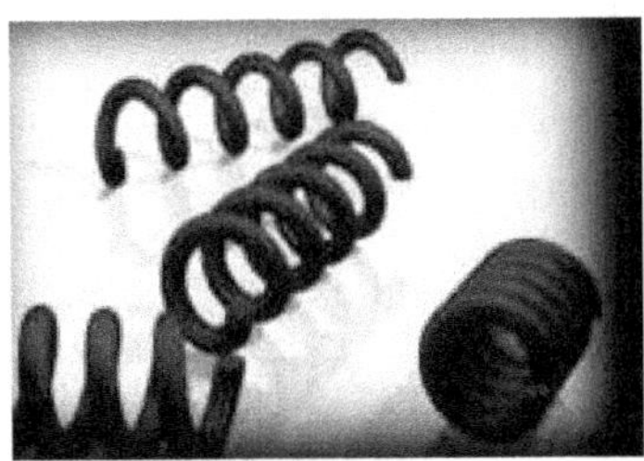

Figure 1.1: Nano Scale Copper

- Shiny orange yellow gold changes its colour to brownish black on reducing the size

1.2.   Scientific Revolutions

The first ever concept was presented in 1959 by the famous professor of physics Dr. Richard Feynman.

There's Plenty of Room at the Bottom: An Invitation to Enter a New Field of Physics" was a lecture given by physicist Richard Feynman at the annual American Physical Society meeting at Caltech on December 29, 1959.

Figure 1.2: Dr Richard Feynman

He also presented the possibility of **"swallowing the doctor**". This concept involved building a tiny, swallowable surgical robot.

The term "**Nano-technology**" had been coined by **Norio Taniguchi in 1974.**

After Feynman had discovered this new field of research catching the interest of many scientists, two approaches have been developed describing the different possibilities for the synthesis of nanostructures. These manufacturing approaches fall under two categories:

- Top-down approach
- Bottom-up approach

Which differ in degrees of quality, speed, and cost.

1.2.1. *Lycurgus Cup*

Nanoparticles and structures have been used by humans in fourth century AD, by the Roman, which demonstrated one of the most interesting examples of nanotechnology in the ancient world. The Lycurgus cup, from the British Museum collection, represents one of the most outstanding achievements in ancient glass industry.

Figure 1.3: Lycurgus cup

The glass appears green in reflected light (Green Colour) and red-purple in transmitted light.

In 1990, the scientists analysed the cup using a transmission electron microscopy (TEM) to explain the phenomenon of dichroism. The observed dichroism (two colours) is due to the presence of nanoparticles with 50–100 nm in diameter.

1.2.2. History of Nano Technology

- 2000 years ago, Sulphide Nano crystals used by Greeks and Romans to dye hair.
- 1000 years age, gold Nano particles of different sizes used to produce different colours in stained glass windows.
- 1959: "There is plenty of room at the bottom" – Richard Feynman.
- 1974: "Nano Technology" Taniguchi named first time.
- 1981: IBM developed Scanning Tunnelling microscope.
- 1985: "Bucky Ball" – scientists at Rice University and University of Sussex discovered C60.

1.2.3. Some of the Discoveries in History

Table 1.1: Discoveries in History

Discovery type	Name	Age	Start Date
Industrial	Tools	Stone	2200000 BC
Industrial	Metallurgy	Bronze	3500 BC
Industrial	Steam power	Industrial	1764
Automation	Mass production	Consumer	1906
Automation	Computing	Information	1946
Health	Genetic Engg	Genetic	1953
Industrial	Nano Tech	Nano Age?	1991
Automation	Molecular assembler	Assembler age?	2020..
All three	Life assemblers	Life age?	2050?

1.3. Nanotechnology and Nanomachines

A molecular nano machine, nanite, or nanomachine is a molecular component that produces quasi-mechanical movements (output) in response to specific stimuli (input). In cellular biology, macromolecular machines frequently perform tasks essential for life, such as DNA replication and ATP synthesis. The expression is often more generally applied to molecules that simply mimic functions that occur at the macroscopic level. The term is also common in nanotechnology where a number of highly complex molecular machines have been proposed that are aimed at the goal of constructing a molecular assembler.

1.3.1. Types of Nano Machines

Molecular machines can be divided into two broad categories;

- Artificial
- Biological.

In general, artificial molecular machines (AMMs) refer to molecules that are artificially designed and synthesized whereas biological molecular machines can commonly be found in nature and have evolved into their forms after abiogenesis on Earth.

Artificial Molecular Machines (AMM)

The artificial molecular machines (AMMs) have been synthesized by chemists which are rather simple and small compared to biological molecular machines. The first AMM, a molecular shuttle, was synthesized by Sir J. Fraser Stoddart. A molecular shuttle is a rotaxane molecule where a ring is mechanically interlocked onto an axle with two bulky stoppers. The ring can move between two binding sites with various stimuli such as light, pH, solvents, and ions. As the authors of this 1991 JACS paper noted: "Insofar as it becomes possible to control the movement of one molecular component with respect to the other in a rotaxane, the technology for building molecular machines will emerge", mechanically interlocked molecular architectures spearheaded AMM design and synthesis as they provide directed molecular motion. Today a wide variety of AMMs exists as listed below.

Molecular Motors

Molecular motors are molecules that are capable of directional rotary motion around a single or double bond. Single bond rotary motors are generally activated by chemical reactions whereas double bond rotary motorshttps://en.wikipedia.org/wiki/Molecular_machine -

cite_note-17 are generally fuelled by light. The rotation speed of the motor can also be tuned by careful molecular design. Carbon nanotube nanomotors have also been produced.

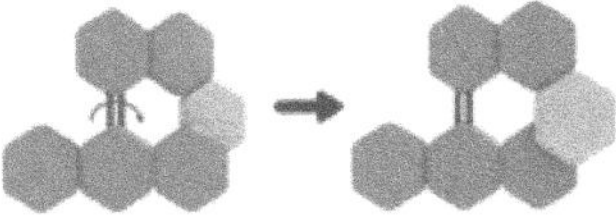

Figure 1.4: Overcrowded Alkane Molecular Motor

Molecular Propeller

A molecular propeller is a molecule that can propel fluids when rotated, due to its special shape that is designed in analogy to macroscopic propellers. It has several molecular-scale blades attached at a certain pitch angle around the circumference of a nanoscale shaft. Also see molecular gyroscope.

Molecular Switch

A molecular switch is a molecule that can be reversibly shifted between two or more stable states. The molecules may be shifted between the states in response to changes in pH, light (photo switch), temperature, an electric current, microenvironment, or the presence of a ligand.

Molecular Shuttle

A molecular shuttle is a molecule capable of shuttling molecules or ions from one location to another. A common molecular shuttle consists of a rotaxane where the macrocycle can move between two sites or stations along the dumbbell backbone.

Figure 1.5: Rotaxane based Molecular Shuttle

Nanocar

Nanocars are single molecule vehicles that resemble macroscopic automobiles and are important for understanding how to control molecular diffusion on surfaces. The first nanocars were synthesized by James M. Tour in 2005. They had an H shaped chassis and 4 molecular wheels (fullerenes) attached to the four corners. In 2011, Ben Feringa and co-workers synthesized the first motorized nanocar which had molecular motors attached to the chassis as rotating wheels. The authors were able to demonstrate directional motion of the nanocar on a copper surface by providing energy from a scanning tunneling microscope tip. Later, in 2017, the world's first ever Nanocar Race took place in Toulouse.

Molecular Balance

A molecular balance is a molecule that can interconvert between two and more conformational or configurational states in response to the dynamic of multiple intra- and intermolecular driving forces, such as hydrogen bonding, solvophobic/hydrophobic effects, π interactions, and steric and dispersion interactions. Molecular balances can be small molecules or macromolecules such as proteins. Cooperatively folded proteins, for example, have been used as molecular balances to measure interaction energies and conformational propensities.

Molecular Tweezers

Molecular tweezers are host molecules capable of holding items between their two arms. The open cavity of the molecular tweezers binds items using non-covalent bonding including hydrogen bonding, metal coordination, hydrophobic forces, van der Waals forces, π interactions, or electrostatic effects. Examples of molecular tweezers have been reported that are constructed from DNA and are considered DNA machines.

Molecular Sensor

A molecular sensor is a molecule that interacts with an analyte to produce a detectable change. Molecular sensors combine molecular recognition with some form of reporter, so the presence of the item can be observed.

Molecular Logic Gate

A molecular logic gate is a molecule that performs a logical operation on one or more logic inputs and produces a single logic output. Unlike a molecular sensor, the molecular logic gate will only output when a particular combination of inputs are present.

Molecular Assembler

A molecular assembler is a molecular machine able to guide chemical reactions by positioning reactive molecules with precision.

Molecular Hinge

A molecular hinge is a molecule that can be selectively switched from one configuration to another in a reversible fashion. Such configurations must have distinguishable geometries; for instance, azobenzene groups in a linear molecule may undergo *cis-trans* isomerization when irradiated with ultraviolet light, triggering a reversible transition to a bent or V-shaped conformation. Molecular hinges typically rotate in a crank-like motion around a rigid axis, such as a double bond or aromatic ring.[53] However, macrocyclic molecular hinges with more clamp-like mechanisms have also been synthesized.

Biological Molecular Machines (BMM)

A ribosome will perform the elongation and membrane targeting stages of protein translation. The ribosome is green and yellow, the tRNAs are dark blue, and the other proteins involved are light blue. The produced peptide is released into the endoplasmic reticulum.

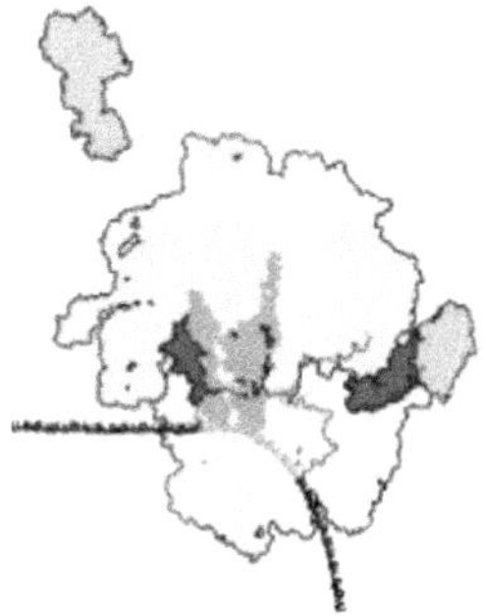

Figure 1.6: Ribosome

The most complex macromolecular machines are found within cells, often in the form of multi-protein complexes. Important examples of biological machines include motor proteins such as myosin, which is responsible for muscle contraction, kinesin, which moves cargo inside cells away from the nucleus along microtubules, and dynein, which moves cargo inside cells towards the nucleus and produces the axonemal beating of motile cilia and flagella.

Still other machines are responsible for gene expression, including DNA polymerases for replicating DNA, RNA polymerases for producing mRNA, the spliceosome for removing introns, and the ribosome for synthesising proteins. These machines and their nanoscale dynamics are far more complex than any molecular machines that have yet been artificially constructed.

1.3.2. Bucky Ball

Buckyballs, also called fullerenes, were one of the first nanoparticles discovered. This discovery happened in 1985 by a trio of researchers working out of Rice University named Richard Smalley, Harry Kroto, and Robert Curl.

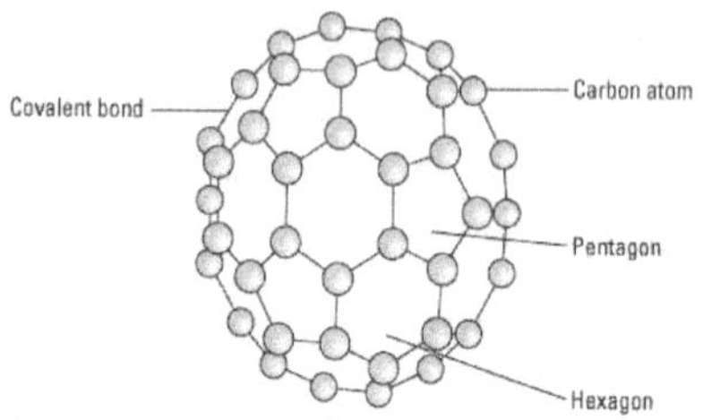

Figure 1.7: Bucky Ball

The covalent bonds between carbon atoms make buckyballs very strong, and the carbon atoms readily form covalent bonds with a variety of other atoms. Buckyballs are used in composites to strengthen material.

Buckyballs have the interesting electrical property of being very good electron acceptors, which means they accept loose electrons from other materials. This feature is useful, for example, in increasing the efficiency of solar cells in transforming sunlight into electricity.

- Bucky balls have ability to behave like "molecular ball bearings" allowing surfaces to glide over one another.
- When metal atoms attached to buckyballs, they can function as catalysts, increasing the speed of chemical reactions.
- They can also be modified to act as super conductors at low temperatures.

1.3.3. Assemblers

During 1980's and 1990's, K Eric Drexler popularised the potential for Nano machines.

The goal of nano machine technology is to build an "assembler" nano machine that is designed to manipulate matter at the atomic level.

The assembler would be able to rearrange atoms from raw materials in order to produce useful items.

Drexler Assembler

Drexler has described the assembler as a device having a nanorobot under computer control. An assembler is a nanomachine, but a very special one that can both build nanomachines and reproduce itself in the same process. It will be capable of holding and positioning atoms and molecules in order to control the precise location at which chemical reactions take place.

Drexler compared **nano machines with ribosomes**.

1.3.4. *Ribosomes*

A typical ribosome is relatively small (a few thousand cubic nanometres) and is capable of building amino acids. Amino acids are organic compounds that combine to form proteins. Amino acids and proteins are the building blocks of life. When proteins are digested or broken down, amino acids are left. The human body uses amino acids to make proteins to help the body, break down food, Repair body tissue, Perform many other body functions

- Ribosomes are macromolecular machines, found within all living cells
- Ribosomes consist of two major components: the small and large ribosomal subunits
- Transfer RNA (tRNA) carries DNA code to the ribosome.
- mRNA molecule carries a portion of the DNA code to other parts of the cell for processing
- A codon is a sequence of three DNA nucleotides that corresponds with a specific amino acid

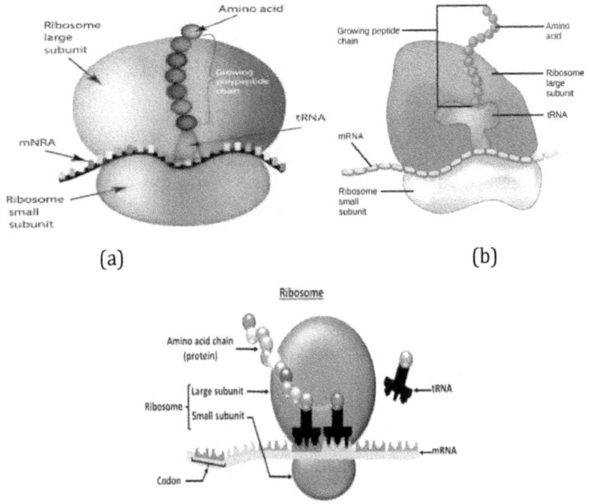

Figure 1.8c: Ribosomes

1.3.5. *Codon*

Codon is a genetic code read in a group of 3 letters

The genetic code is the set of rules by which information encoded in genetic material (DNA or RNA sequences) is translated into proteins by living cells.

Figure 1.9: CODON

1.4. Atomic Structure & The Periodic Table

1.4.1. Atomic Structure

Ancient Greeks developed the concept of elements as basic substances from which all forms of matter can be built. It was believed that when four major elements; earth, fire, air and water, were combined in correct proportions, they could produce all of the other substances.

In 1807, English schoolteacher, John Dalton (1766–1844) revived the concept of atoms and proposed an atomic theory on facts and experimental evidence.

- Elements are composed of indivisible particles called atoms.
- All atoms of a given element are identical (now known to be incorrect); the atoms of different elements are different in some fundamental way.
- Chemical compounds are formed when atoms combine with each other. A given compound always has the same relative number and types of atoms.
- Chemical reactions involve reshuffling of atoms from one set of combinations to another. The atoms themselves are not changed in a chemical reaction.

Dalton considered the atom to be indivisible, which is not true. Indivisible' means that atoms cannot be broken down further without changing the chemical nature of the element. For example, when a carbon atom is broken down into smaller particles (known as subatomic particles), it loses its chemical properties.

There are three major fundamental subatomic particles: electrons, protons and neutrons. The protons and neutrons are collectively known as nucleons.

Electrons are tiny, very light particles that have a negative charge (-). Protons are much larger and heavier than electrons and have a positive charge (+). Neutrons are large and heavy like protons but have no charge.

Since the total charge on atoms is neutral, they must have equal numbers of electrons and protons. All isotopes of a given element have the same number of protons but different numbers of neutrons in each atom.

The structure of the nucleus can mean that an atom can behave like a small magnet, and in a magnetic field it can be aligned either with or against the magnetic field. This magnetic property of the nucleus is called its magnetic moment. Only some nuclei have magnetic moments and this depends on the ratio of protons to neutrons hydrogen (one proton) with 1, 2 and 3 neutrons respectively

$$_1^1H, \ _1^2H, \ _1^3H \qquad _2^4He, \ _3^6Li, \ _6^{12}C$$

If there is an equal number of protons and neutrons then, the nucleus does not have a magnetic moment. This can be important in detecting these nuclei using their magnetic fields. The phenomenon is known as nuclear magnetic resonance (NMR).

Some types of isotopes are known as radioisotopes, which undergo spontaneous nuclear changes that cause them to be transformed into other elements. It was Henry Becquerel who discovered that atoms of some elements are not stable and disintegrate naturally. Uranium was the first element to be found to be unstable. In natural radiation, atoms spontaneously emit radiation of various types, known as alpha, beta and gamma particles.

This phenomenon was later called radioactivity by Marie Curie, and was also found to occur in the element's polonium and radium. The existence of radioisotopes has had practical applications, such as in carbon dating and in chemotherapy for thyroid cancer.

In 1911, Ernest Rutherford, a New Zealander, provided one of the most significant developments in the understanding of the structure of an atom. Prior to this time, it was thought that the atom had a uniform density.

In Rutherford's experiment to test this model, he directed particles, which carry a positive charge, at a thin sheet of metal foil. If the previous model was accurate the particles should either pass through the foil virtually undisturbed or all be deflected back at 360 degrees.

1.4.2. The Periodic Table

The periodic table is made up of rows and columns.

- An element is identified by its chemical symbol.
- The number above the symbol is the atomic number
- The number below the symbol is the rounded atomic weight of the element.
- A row is called a period
- A column is called a group

The periodic table is an arrangement of the atoms (called elements) that groups them according to similar properties.

Atomic Number: The atomic number or proton number of a chemical element is the number of protons found in the nucleus of every atom of that element.

Mass Number: it is the number of protons and neutrons

The periodic table organizes elements according to similar properties so you can tell the characteristics of an element just by looking at its location on the table.

In 1869, a Russian chemist and teacher published table of the elements.

Mendeleev arranged the elements in the periodic table in order of increasing atomic mass.

In 1913, **Henry Moseley** through his work with X-rays, determined the atomic number of the elements. He rearranged the elements in order of increasing atomic number.

Periodic Table of the Elements

Key (example): 11 = Atomic number, **Na** = Element symbol, Sodium = Element name, 22.990 = Atomic weight

Legend: Alkalai metals, Alkaline earth metals, Lanthanides, Actinides, Transition metals, Unknown properties, Post-transition metals, Metalloids, Other nonmetals, Halogens, Noble gases

Period	1 / 1A	2 / 2A	3 / 3B	4 / 4B	5 / 5B	6 / 6B	7 / 7B	8 / 8B	9 / 8B	10	11 / 1B	12 / 2B	13 / 3A	14 / 4A	15 / 5A	16 / 6A	17 / 7A	18 / 8A
1	1 H Hydrogen 1.0078																	2 He Helium 4.0026
2	3 Li Lithium 6.938	4 Be Beryllium 9.0122											5 B Boron 10.806	6 C Carbon 12.009	7 N Nitrogen 14.806	8 O Oxygen 15.999	9 F Fluorine 18.998	10 Ne Neon 20.180
3	11 Na Sodium 22.990	12 Mg Magnesium 24.305											13 Al Aluminum 26.982	14 Si Silicon 28.084	15 P Phosphorus 30.974	16 S Sulfur 32.059	17 Cl Chlorine 35.446	18 Ar Argon 39.948
4	19 K Potassium 39.098	20 Ca Calcium 40.078	21 Sc Scandium 44.956	22 Ti Titanium 47.867	23 V Vanadium 50.942	24 Cr Chromium 51.996	25 Mn Manganese 54.938	26 Fe Iron 55.845	27 Co Cobalt 58.933	28 Ni Nickel 58.693	29 Cu Copper 63.546	30 Zn Zinc 65.38	31 Ga Gallium 69.723	32 Ge Germanium 72.63	33 As Arsenic 74.922	34 Se Selenium 78.96	35 Br Bromine 79.904	36 Kr Krypton 83.798
5	37 Rb Rubidium 85.468	38 Sr Strontium 87.62	39 Y Yttrium 88.906	40 Zr Zirconium 91.224	41 Nb Niobium 92.906	42 Mo Molybdenum 95.96	43 Tc Technetium 98.9062	44 Ru Ruthenium 101.07	45 Rh Rhodium 102.91	46 Pd Palladium 106.42	47 Ag Silver 107.87	48 Cd Cadmium 112.41	49 In Indium 114.82	50 Sn Tin 118.71	51 Sb Antimony 121.76	52 Te Tellurium 127.60	53 I Iodine 126.90	54 Xe Xenon 131.29
6	55 Cs Cesium 132.91	56 Ba Barium 137.33	57–71 (Lanthanides)	72 Hf Hafnium 178.49	73 Ta Tantalum 180.95	74 W Tungsten 183.84	75 Re Rhenium 186.21	76 Os Osmium 190.23	77 Ir Iridium 192.22	78 Pt Platinum 195.08	79 Au Gold 196.97	80 Hg Mercury 200.59	81 Tl Thallium 204.38	82 Pb Lead 207.2	83 Bi Bismuth 208.98	84 Po Polonium (209)	85 At Astatine (210)	86 Rn Radon (222)
7	87 Fr Francium (223)	88 Ra Radium (226)	89–103 (Actinides)	104 Rf Rutherfordium (261)	105 Db Dubnium (262)	106 Sg Seaborgium (266)	107 Bh Bohrium (264)	108 Hs Hassium (269)	109 Mt Meitnerium (268)	110 Ds Damstadtium (268)	111 Rg Roentgenium (268)	112 Cn Copernicium (268)	113 Uut Ununtrium (268)	114 Fl Flerovium (268)	115 Uup Ununpentium (268)	116 Lv Livermorium (268)	117 Uus Ununseptium (268)	118 Uuo Ununoctium (268)

Lanthanides:

57 La Lanthanum 138.91	58 Ce Cerium 140.12	59 Pr Praseodymium 140.91	60 Nd Neodymium 144.24	61 Pm Promethium (145)	62 Sm Samarium 150.36	63 Eu Europium 151.96	64 Gd Gadolinium 157.25	65 Tb Terbium 158.93	66 Dy Dysprosium 162.50	67 Ho Holmium 164.93	68 Er Erbium 167.26	69 Tm Thulium 168.93	70 Yb Ytterbium 173.04	71 Lu Lutetium 174.97

Actinides:

89 Ac Actinium (227)	90 Th Thorium 232.04	91 Pa Protactinium 231.04	92 U Uranium 238.05	93 Np Neptunium (257)	94 Pu Plutonium (244)	95 Am Americium (243)	96 Cm Curium (247)	97 Bk Berkelium (247)	98 Cf Californium (251)	99 Es Einsteinium (252)	100 Fm Fermium (257)	101 Md Mendelevium (258)	102 No Nobelium (259)	103 Lr Lawrencium (262)

Figure 1.10: The Periodic Table

In the modern periodic table elements are arranged in order of increasing atomic number. Periodic Law states: When elements are arranged in order of increasing atomic number, there is a periodic repetition of their physical and chemical properties.

Three classes of elements are

- Metals,
- Nonmetals, and
- Metalloids.

Across a period, the properties of elements become less metallic and more non-metallic.

Properties of Metals

- Metals are good conductors of heat and electricity.
- Metals are shiny.
- Metals are ductile (can be stretched)
- Metals are malleable (can be poured into thin sheets)
- A chemical property of metal is its reaction with water which results in corrosion.
- Solid at room temperature except Hg.

Properties of Non-metals

- Poor conductors of heat and electricity
- Non ductile
- Non malleable
- Solid non metals are brittle and break easy.
- They are dull
- Many non-metals are gases.

Properties of Metalloids

- Have properties like both metals and non-metals.
- They are solids that can be shiny or dull.
- They conduct heat and electricity better than non-metals nut not as well as metals
- They are ductile and malleable.

There are **4 chemical groups** in Periodic table.

- Alkali Metals (IA)   :   H, Li, Na, K, Rb, Cs & Fr
- Alkaline earth metals (IIA)   :   Be, Mg, Ca, Sr, Ba & Ra
- Halogens (VII)   :   F, Cl, Br, I & At
- Noble gases (VIIIA)   :   He, Ne, Ar, Kr, Xe & Rn

Alkali Metals

- The alkali family is found in first column of periodic table.
- Atoms of alkali metals have a single electron in their outermost level, in other words, 1 valance e⁻
- They are shiny, have the consistency of clay and are easily cut with a knife.
- They are most reactive metals
- They react violently with water.
- Alkali metals are never found as free elements in nature. They are always bounded with another element.

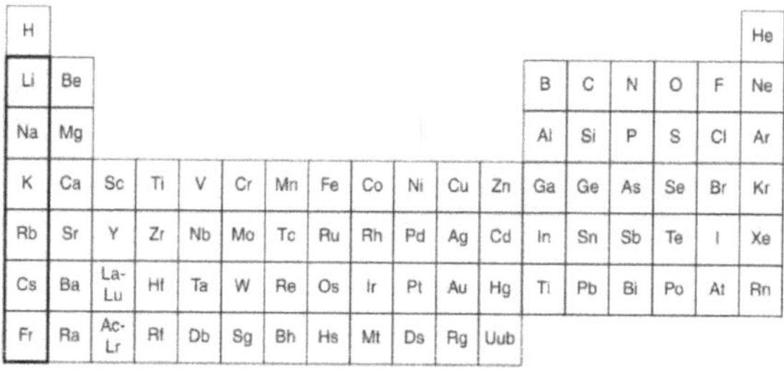

Figure 1.11: Alkali Metals in Periodic Table

Alkaline Earth Metals

- They are never found uncombine in nature.
- They have two valance electrons.
- Alkaline earth metals include magnesium and calcium among others.
- Second column of periodic table.

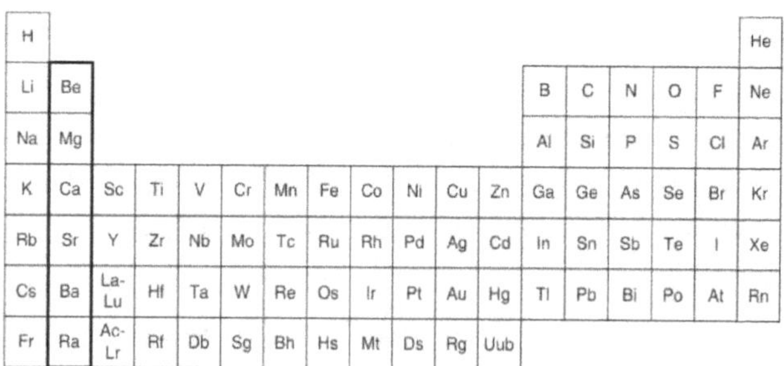

Figure 1.12: Alkali Earth Metals

Transition Elements

- These are the most common metal: copper, tin, zinc, iron, nickel, gold, and silver.

- They are good conductors of heat and electricity.

- The compounds of transition metals are usually brightly colored and are often used to color paints. Transition elements have 1 or 2 valence electrons, which they lose when they form bonds with other atoms.

- Transition metals combine chemically with oxygen to form compounds called oxides.

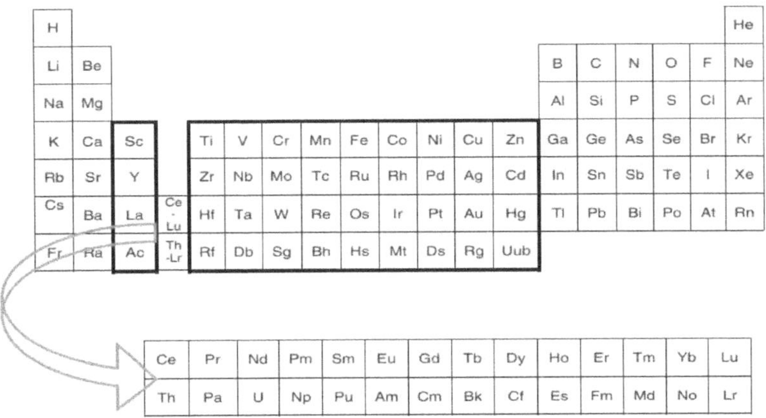

Figure 1.13: Transition Elements

Noble Gases

- On the far-right hand side of the periodic table are the noble gases.

- These combine with very few elements. They are therefore very unreactive and make useful materials for developing nanotechnological tools, especially the largest atoms such as radon, Rn and xenon, Xe.

- All the noble gases have full outer shells. The list includes helium, neon (Ne), argon (Ar), krypton (Kr), xenon (Xe), and radon (Rn)

- Neon is used in advertising signs. Argon is used in light bulbs. Helium is used in balloons and to cool things. Xenon is used in headlights for new cars.

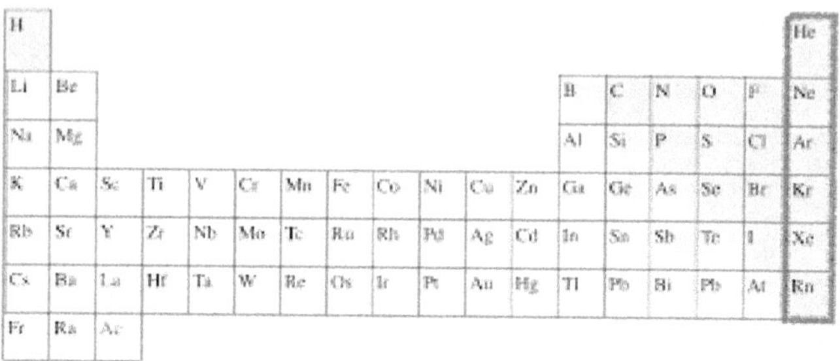

Figure 1.14: Noble Gases

Halogen Group

- They have seven elements in their outer shell
- The elements included are fluorine (F), chlorine (Cl), bromine (Br), iodine (I), and astatine (At).
- They are very reactive.
- Fluorine is the most **reactive** and combines with most elements from around the periodic table. Reactivity decreases as you move down the column.

Atomic Radius

The atomic radius of a chemical element is a measure of the size of its atoms, usually the typical distance from the centre of the nucleus to the outermost shell of electrons or mean distance between two centres of the nucleus

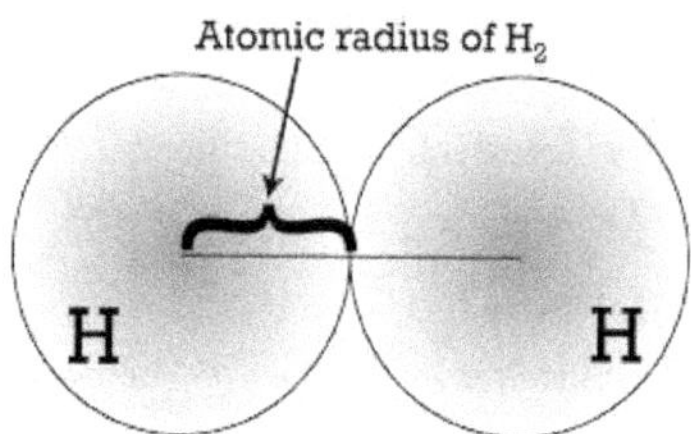

Figure 1.15: Atomic Radius

- As you go down a column, atomic radius increases. Attraction between nucleus and electrons are less
- As we move right side in the period, atomic radius decreases.
- Attraction between nucleus and electrons are strong

Ionic Radius

Ions are formed when an atom loses or gains electrons. When an atom loses an electron, it forms a cation and when it gains an electron it becomes an anion. The Ionic radius can be described as the distance between the centres of cation and anion

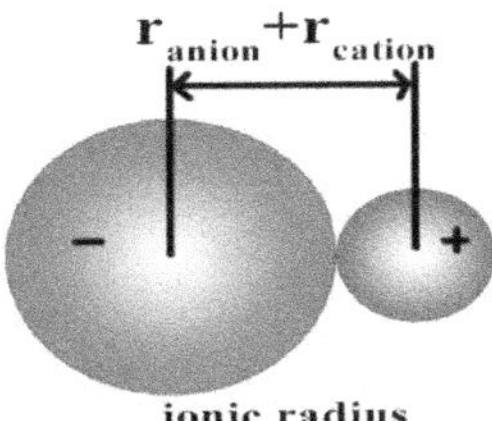

Figure 1.16: Ionic Radius

- As you move across a row of the periodic table, the ionic radius decreases for metals forming cations, as the metals lose their outer electron orbitals.
- As you move across column of the periodic table, the ionic radius increases for non-metals forming anions

1.5. Molecules and Phases, Energy

Nano Science: Nanoscience is the study of structures and materials on an ultra-small scale, and the unique and interesting properties these materials demonstrate. Nanoscience is cross disciplinary, meaning scientists from a range of fields including chemistry, physics, biology, medicine, computing, materials science and engineering are studying it and using it to better understand our world.

1.5.1. Molecules and Phases

Phases are states that we define by their properties, such as liquids, gases and solids. Molecules are collections of atoms bound to each other that exist in these phases.

Oxygen that we breathe is a molecule made up of two oxygen atoms combined, and written as O_2. Likewise, water, H_2O, is the combination of two hydrogen atoms and one oxygen atom.

Molecules are formed by atoms in one of two ways. One way is by Sharing electrons. This type of combination is called a covalent bond. Secondly, molecules can be formed by transferring electrons between each other, which is called Ionic bond

When an electron is transferred between two atoms, they are said to be ionised. An example is sodium chloride, NaCl, in which the sodium atom is positive, Na+; and the chlorine atom is negative, Cl–. A positive ion is called a cation and a negative ion is called an anion.

When there is a covalent bond, as in chlorine, the bond is written as a straight-line Cl–Cl. The line represents the two shared electrons.

In covalent solids, the molecules are discrete and can be differentiated from each other. In ionic solids such as sodium chloride the atoms are arranged in a regular pattern in space called a lattice

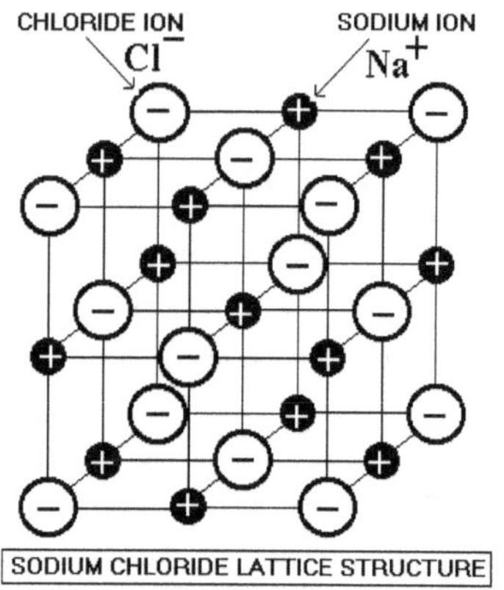

Figure 1.17: Sodium Chloride Lattice Structure

The large spheres are chloride ions, and the small spheres are sodium ions. The lines between the atoms show the bonds that hold the structure together.

In metals all the electrons are shared by all the atoms at one time. Thus metals easily conduct electricity because an extra electron can be added or removed without removing it from a single discrete atom.

There are some other types of bonding which are important in nanotechnology. Molecules and atoms can also be drawn together by relatively weak forces collectively called **Vander Waals forces** These types of force are named after Johannes's van derwaals who studied them in gases. There are three types of these forces, called as dispersive forces; dipole-dipole forces; and hydrogen bonding.

These forces are responsible for the liquids, solids and solutions state of any compound.

The order of strength of these intermolecular forces is given below.

London's dispersion force < dipole-dipole < H-bonding

Vander Waals Forces

In molecular physics, the Van der Waals force, named after Dutch physicist Johannes Diderik van der Waals, is a distance-dependent interaction between atoms or molecules. Unlike ionic or covalent bonds, these attractions do not result from a chemical electronic bond; they are comparatively weak and therefore more susceptible to disturbance. The Van der Waals force quickly vanishes at longer distances between interacting molecules.

Van der Waals force plays a fundamental role in fields as diverse as supramolecular chemistry, structural biology, polymer science, nanotechnology and surface science

London dispersion forces are a type of force acting between atoms and molecules that are normally electrically symmetric; that is, the electrons are symmetrically distributed with respect to the nucleus.

Polar Molecules: A polar molecule is usually formed when the one end of the molecule is said to possess more number of positive charges and whereas the opposite end of the molecule has less negative charges, creating an electrical pole. When a molecule is said to have a polar bond, then the centre of the negative charge will be one side, whereas the centre of positive charge will be in the different side. The entire molecule will be a polar molecule.

Non- Polar Molecules: A molecule which does not have the charges present at the end due to the reason that electrons are finely distributed and those which symmetrically cancel out each other are the non- polar molecules. In a solution, a polar molecule cannot be mixed with the non-polar molecule. For example, consider water and oil. In this solution, water is a polar molecule whereas oil behaves as a non-polar molecule. These two molecules do not form a solution as they cannot be mixed up.

Dipole, literally, means "two poles," two electrical charges, one negative and one positive. Dipoles are common in atoms whenever electrons (-) are unevenly distributed around nuclei (+), and in molecules whenever electrons are unevenly shared between two atoms in a covalent bond.

Dipole-dipole forces are attractive forces between the positive end of one polar molecule and the negative end of another polar molecule.

Hydrogen bonding is a type of dipole-dipole interaction between molecules

It results from the attractive force between a hydrogen atom covalently bonded to a high electronegative atom such as a N, O, or F atom The electrons move about a nucleus a bit like a wave on the sea. In the sea there is more water in the bulk of the wave than at the apex at any point of time in space, but this changes as the wave moves. In an atom, the electrons tend to 'wash' around the nucleus like a wave, and more electrons can be found at any point of time in a given space. This creates charge differences in space. This charge separation is known as a dipole moment.

1.5.2. *Energy*

- protons and neutrons are made of matter which is also just energy.
- Light is also a form of energy that we call electromagnetic radiation. It can be described either as a wave, or as consisting of matter called photons.
- Distance between Crests is called wavelength

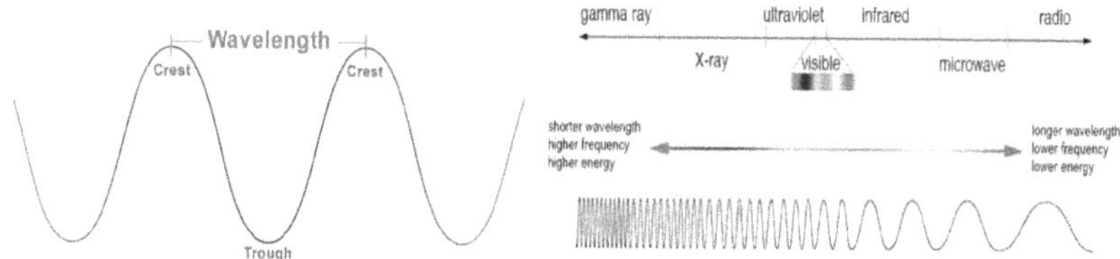

Figure 1.18: Gamma Radiation

Gamma rays have the smallest wavelengths and the most energy of any wave in the electromagnetic spectrum. They are produced by the hottest and most Energetic objects in the universe, such as neutron stars and pulsars, supernova explosions. On Earth, gamma waves are generated by nuclear explosions and lightning

- Gamma rays are emitted by the nucleus and A material containing unstable nucleus is considered radioactive.
- Radioactive decay (also known as nuclear decay, radioactivity, radioactive disintegration or nuclear disintegration) is the process by which an unstable atomic nucleus loses energy by radiation.
- Gamma rays are a product of gamma decay: how the nucleus gets rid of excess energy. In gamma decay the nucleus emits a high-energy photon (electromagnetic radiation), gamma rays, but the number of protons and neutrons stays the same.

- All electromagnetic radiation flows through empty space at 3×10^8 metres per second, about 670 million miles per hour.
- electromagnetic radiation is a kind of radiation including visible light, radio waves, gamma rays, and X-rays, in which electric and magnetic fields vary simultaneously.

Clearly, as the wavelength shortens the frequency also increases. When the frequency is extremely high, it becomes continuous. Then, waves do not exist. the matter is present continuously, and the energy is better described as mass. Indeed, some forms of energy have properties of both mass and waves. The duality of energy as mass or wave is important in nanotechnology, since at the atomic level these materials may interact, exhibiting any of the properties of the two states.

The de Broglie equation expresses the relationship between wavelength λ, and mass as:

De Broglie's Equation

$$\lambda = \frac{h}{mv}$$

Where
λ = wavelength in meters
v = the velocity in meters/sec
m = the mass in kilograms
h = Plancks's constant in J/Hz

According to this equation, a heavy particle travelling slowly could have the same wavelength as a light particle travelling quickly.

Quantization is the process of constraining an input from continuous values to a discrete value. When atoms have been studied by exciting the electrons to a higher energy state and then observed, it is clear that they lose energy in only discrete amounts. This phenomenon is also quantisation.

The energy is quantised and the electron is also said to be quantised. Thus, the electron, when described as a wave, can only have certain wavelengths.

1.5.3. Enthalpy and Entropy

- Enthalpy (H) is the heat content of a system at constant pressure.
- The heat that is absorbed or released by a reaction at constant pressure is the same as the enthalpy change, and is given the symbol ΔH
- Entropy is a measure of how much energy is not available to do work
- Both Entropy and Enthalpy must be considered while manipulating atoms.

1.6.   Surfaces and Dimensional Shape

Importance of Dimension

- For one dimensional object, they would be a line or a dot of infinitely small thinness.
- For two dimensional objects, they would be flat like a square and also of infinitely small thinness.
- For three dimensions, there are objects that can be observed like sphere, a box, human beings etc

By understanding how three-dimensional objects are represented in two dimensions on paper.

We draw them in perspective as a cube. A box can be easily represented by additional lines to depict the third dimension. Similarly, we can draw four dimensional objects from three dimensions.

For understanding Nano materials, four dimensional will be useful. Imagine if you were a two-dimensional object living in a two-dimensional world but in reality, there were three dimensions. It would be possible to disappear without breaking the laws of physics by moving into the third dimension. These concepts are important in nanotechnology because as objects get smaller, their dimensions become smaller and smaller.

It is possible that in nanotechnology we may discover objects that are better described as one- or two-dimensional, or even four-dimensional.

Indeed, researchers looking for additional dimensions have suggested that to account for changes in the speed of light in space, there must be other dimensions and that they will be discovered on looking closer at atoms.

Importance of Surface Area

Surface area to volume ratio increases as the radius of the sphere decreases and vice versa. Therefore, materials made of nanoparticles have a much greater surface area per unit volume ratio compared with the materials made up of bigger particles The surface areas of nanoparticles are responsible for some properties at the nanoscale. As size decreases the surface area increases. For example, a packet of sugar, which has many small particles of sugar, would dissolve in a solution faster than a sugar cube, or large particle of sugar, because the packet of sugar particles has a greater surface area than the cube of sugar.

Substances at the nanoscale level have a greater surface-to-volume ratio, which causes them to react very quickly. Small particles have a greater percentage of atoms on their surface, which accounts for the increased surface to volume ratio.

1.7. Top Down and Bottom-up Approach

When the size of a material is reduced from a large or macroscopic size, such as meter or a centimetre, to a very small size, the properties remain the same at first. Then changes begin to occur until finally when size drops below 100nm, dramatic changes in properties can occur.

If one dimension is reduced to the nano range, while the other 2 dimensions remains large, then that structure is called as a **"Quantum Well"**.

If two dimensions reduced, and one remain unchanged, structure is known as "**Quantum Wire**".

If all the three dimensions reach the low nano meter range, then it is called as "**Quantum Dot**"

The word "Quantum" is associated with these three types of nano structures, because the changes in properties arise from quantum mechanical nature of physics in the domain of ultrasmall.

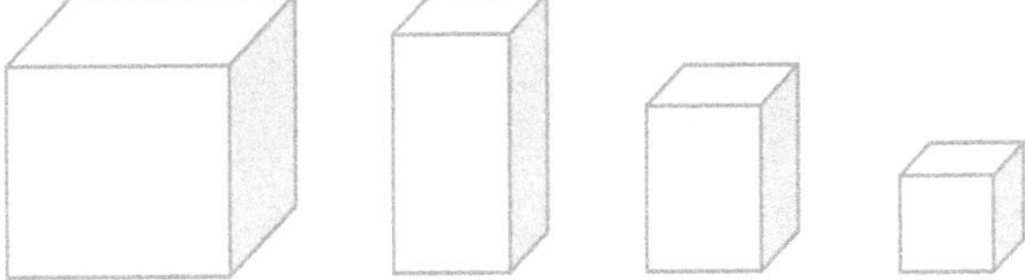

Figure 1.19: Progressive Generation of Rectilinear Geometry (a) Bulk (b) Well (c) Wire (d) Dot

The above process of diminishing the size for the case of rectilinear geometry.

For the "Curvilinear geometry", we can reduce the size or dimensions in circular shape to very small.

Figure 1.20: Progressive Generation of Curvilinear Geometry (a) Bulk (b) Well (c) Wire (d) Dot

While doing these processes, the electronic properties are changed by doing alterations / dimensions.

Field of quantum nano structures have been survived by Dr. Jack et al in 1998.

Preparation of quantum nano structures will be done by 2 types of approaches.

- Bottom-up approach
- Top-down approach

Bottom – Up Approach

In "Bottom-up" approach, small molecules or even atoms are assembled to form the required structure. All the Bottom-up techniques, the starting material is either gaseous state or liquid state of matter

$$10^{-9} \ to \ 10^{-1}$$

- Bottom-up approach refers to the build-up of a material from the bottom: atom by atom, molecule by molecule
- Atom by atom deposition leads to formation of Self- assembly of atoms/molecules and clusters
- Bottom-up refers to methods where devices 'create themselves' by self-assembly. Chemical synthesis is a good example. Bottom-up should broadly speaking be able to produce devices much cheaper than top-down methods, but getting control over the methods is difficult when things become larger and bulkier than what is normally made by chemical synthesis.

It is to collect, consolidate and fashion individual atoms and molecules in to structure. This is carried out by a sequence of chemical reactions controlled by catalysts.

Catalyst: A substance that increases the rate of a chemical reaction without itself undergoing any permanent chemical change.

Ex: zeolites, nickel, platinum, and palladium

This type of approach widely used in Biology. Where, for ex. Catalysts called enzymes assembles amino acids to construct using tissue that forms and supports the organs of body.

Top-Down Approach

This is an opposite approach to bottom-up, which starts with a large-scale object or pattern and gradually reduces its dimension.

- Top-down approach refers to slicing or successive cutting of a bulk material to get nano sized particle
- In Top-down techniques, the starting material is solid state
- Top-down refers to the traditional workshop or microfabrication method where tools are used to cut, mill and shape materials into the desired shape and order.

$$10^{-1} \; to \; 10^{-9}$$

This can be done by a technique called "Lithography" which shines radiation through a template on to a surface coated with a radiation sensitive resist. Then the resist removed and the surface is chemically treated to produce the nano structures.

A typical resist material is the polymer polymethyl methacrylate $[C_5O_2H_8]$

With a molecular weight in the range from 10^5 to 10^6 da.   [da = Dalton]

The lithographic process can be illustrated by starting with a square quantum well located on a substrate. The final product to be produced from the material [ex. GaAs] of quantum well is either a quantum wire or quantum dot.

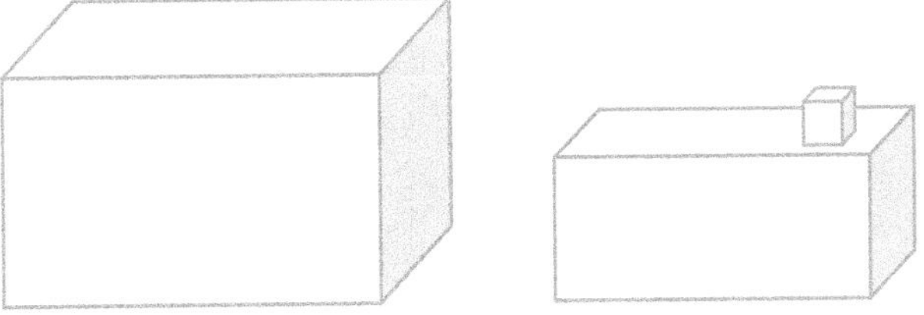

Figure 1.21: (a) Gallium Arsenide Quantum Well on a Substrate, (b) Quantum Well and Quantum dot Formed by Lithography

The steps to be followed in Lithographic process by electron beam lithography.

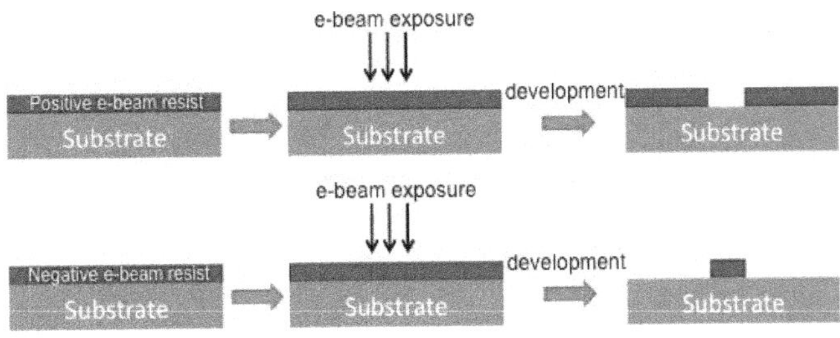

Figure 1.22: Steps Involved in Lithographic Process

(a) Initial quantum well on a substrate and covered by resist

(b) Radiation with sample shielded by template

(c) Configuration after dissolving irradiated portion of resist by developer.

(d) Disposition after addition of etching mask

(e) Arrangement after removal of remainder of resist

(f) Configuration after etching away the unwanted quantum well material

(g) Final nano structure on substrate after removal of etching mask.

Substrate: substrate is the chemical species being observed in a chemical reaction, which reacts with a reagent to generate a product.

It can also refer to a surface on which other chemical reactions are performed. Or play a supporting role in a variety of spectroscopic and microscopic techniques.

Steps in Lithographic Process

- Step 1: Place a radiation sensitive resist on the surface, of the sample substrate.

- Step 2: The sample is irradiated by an electronic beam in the region where the nano structure will be located.
  - o Irradiation can be done by either radiation mask that contains nano structure pattern, or a scanning election beam that strikes the surface only in the desired region.
  - o The radiation chemically modifies the exposed area of the resist so that it becomes soluble in a developer.

- Step 3: The application of developer to remove the irradiated portion of the resist.

- Step 4: The insertion of an etching mask into the hole in the resist.

- Step 5: lifting off the remaining parts of resist.

- Step 6: The areas of quantum well not covered by etching mask are chemically etched away to produce the quantum structure covered by etching mask.

- Step 7: finally etching mask is removed, if necessary to provide the desired quantum structure. Which might be the quantum wire or quantum dot.

Electron beam lithography, which makes use of an electron beam for radiation, other types of lithography employ neutral atom beams (ex: LI, Na, K, Rb & Cs), charged ion beams, or electromagnetic radiation such as visible light, uv light and x rays.

When laser beams are utilised, frequency doublers and quadruples can bring the wave length into a range of $\lambda \sim 150nm$ that is convenient for quantum dot fabrication.

Photo chemical etching can be applied to a surface activated by laser light.

The lithographic technique can be used to make more complex quantum structures then the quantum wire and quantum dot structures.

Also, these lithographic techniques used for multiple quantum well structures.

Advantages of Top-down Approach

- Large scale production: deposition over a large substrate is possible
- Chemical purification is not required.

Disadvantages of Top-down Approach

- Expensive technique
- Broad size distribution (10-1000 nm)
- Varied particle shape or geometry
- Control over disposition parameters is difficult
- Impurities: stress, defects and imperfections get introduced.

Advantages of Bottom-up Approach

- Ultrafine nano particles, nano shells, nano tubes can be prepared
- Deposition parameters can be controlled.
- Narrow size distribution is possible (1-20nm)
- Cheaper technique

Disadvantages of Top-down Approach

- Large scale production is not possible
- Chemical purification of nano particles is required.

Unit - II

Molecular Nano Technology

In this chapter, we will know that how we 'see' atoms using atomic microscopy. And will find out how we can pick up atoms and move them around, or spray them onto surfaces to form nanomaterials.

2.1. Atoms by Inference

The atomic theory predicts that atoms combine together in whole numbers but not in fractions. The chemical bond is associated with the concept of valency. Valency is the combining capacity of atoms.

Chemical bond refers to the formation of a bond chemically between 2 or more atoms, molecules, or ions to give rise to a chemical compound.

A chemical bond is an attraction between atoms. This attraction may be seen as the result of different behaviours of the outermost or valence electrons.

2.1.1. Octet Rule

Atoms tend to gain or lose or share electrons until they are surrounded by 8 valence electrons.

The valency is the number of the electrons that an atom gains, loses or even shares during a chemical reaction.

The valency is the number of electrons that an atom needs to gain or lose in order to achieve noble gas electronic configuration (8 Valence electrons).

If Ratio of atoms in the product is certain then the weight ratio is known.

For example, fluorine and hydrogen combine in the ratio of 19 to 1: that is, 1 gram of hydrogen combines with 19 grams of fluorine. Since the product is HF, which contains one fluorine atom and one hydrogen atom, it follows that the 19_F fluorine atom is 19 times heavier than hydrogen.

For hydrogen, number of atoms in a gram is believed to be about 6.02×10^{23} and is called Avogadro's number, in honour of a famous scientist who did much of the pioneering work on the ratios in which atoms combine.

Avogadro's number of atoms is called a mole. Thus, two moles contain 12.04×10^{23} atoms.

One atom of 1H hydrogen weighs 0.166×10^{-23} grams.

Therefore, one mole weighs $0.166 \times 10^{-23} \times 6.02 \times 10^{23} = 1\text{gram}$.

If we want to weigh out atoms in quantities that can be measured in the laboratory, we can weigh them out in grams and express the quantity of atoms in moles.

This allows us to make sure we have nearly the right number of atoms of each reactant when undertaking a chemical reaction to make a product.

Recently tools have been developed that allow us to see atoms. It is therefore now possible in theory to measure Avogadro's number. This means that for the first time we can count actual atoms and in principle we could count out a mole, but we would be counting for a long time — about 10 000 billion years if we counted at our normal rate of a few units per second, 1, 2, 3 etc.

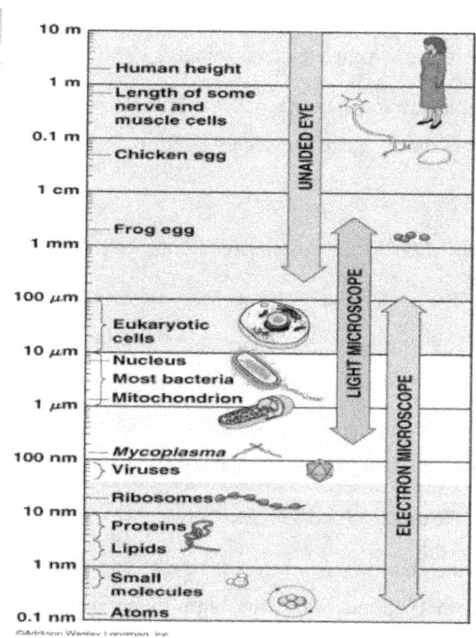

Figure 2.1: Different Sizes Explained by Atom Inference

2.2. Electron Microscope

An electron microscope is a microscope that uses a beam of scattered electrons as a source of illumination.

The electron microscope uses a beam of electrons to magnify an object's image, unlike the optical microscope that uses visible light to magnify images.

The electron microscope uses electron beams and magnetic fields to produce the image instead of light waves and glass lenses used in the light microscopes.

Comparison between Electron and Light Microscope

Table 2.1: Comparison between Electron and Light Microscope

Electron Microscope	Light Microscope
Advantages:	**Advantages:**
• Greater Magnification up to 1000000 X	• Magnification only up to 1000 – 1500x
• Greater depth of field (with SEM, 3D)	• Less depth of field
• No complex lighting	• May involve complex lighting requirements
• Greater resolution up to 1nm	• Less resolution up to 1 micron
Disadvantages:	**Disadvantages:**
• Expensive (1 million)	• Cheap
• Expensive to run	• Cheaper to run
• Image black and white	• Image colour
• Dead specimen	• Live or dead specimen
• Requires vacuum	• No vacuum required
• Expert preparation	• Simple preparation
• Special large room	• Small and portable
• Affected by magnetic fields.	• Unaffected by magnetic fields.

The stepwise stages in obtaining an image in a electron microscope are:

- A stream of electrons is released by an electron source and accelerated toward the specimen using a positive electrode potential
- Stream of electrons is confined and focused using metal apertures and magnetic lenses into a thin, focused, monochromatic beam.
- The beam is focused onto the sample using a magnetic lens
- Interactions occur inside the irradiated sample, affecting the electron beam.
- These interactions and effects are detected and transformed into an image.

2.2.1. *Why We Need Electron Microscope*

- Light microscopes are limited by the physics of light to 500x or 1000x magnification and a resolution of 0.2 micrometres.

- In the early 1930, there was a scientific desire to see the fine details of the interior structures of organic cells.
- This requires 10000 x plus magnification which was just not possible using light microscopic. So, we need electronic microscope.

2.3. Classification of Electron Microscopes

There are 2 types of electron microscopes.

- Scanning Electron Microscope (SEM)
- Transmission Electron Microscope (TEM)

2.3.1. *Scanning Electron Microscope (SEM)*

- Images produced by scanning the surface with a focused beam of electrons
- Electrons interact with the atoms of the sample resulting various signals containing information about surface topography, composition and other properties
- Electron beam scanned in a raster scan pattern: Resolution < 1 nm
- Works in a wide range of pressure and temperature

History of SEM

- In 1935, the original prototype of the SEM, was made by Knoll(Germany).
- In 1942, Zworykin (USA), developed a SEM for observing a bulk specimens.
- In 1965, Cambridge Scientific Instrument (UK) & JOEL (Japan) first commercialized SEM individually.

Components of SEM

- Electronic Gun
- Condenser Lenses
- Objective Aperture
- Objective Lenses
- Scan Coils
- Chamber
- Detectors
- Image display unit

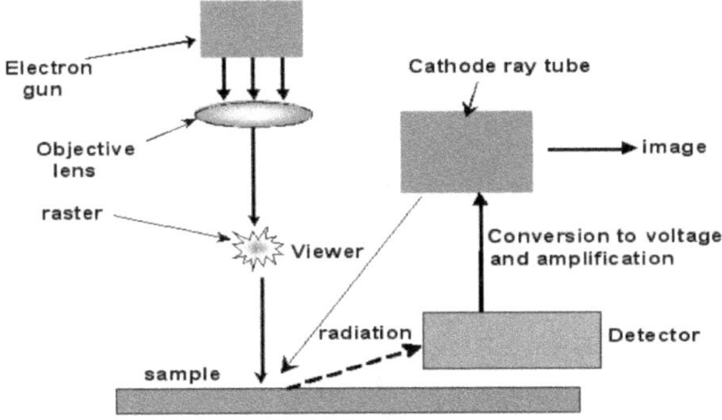

Figure 2.2: Block Diagram of Scanning Electron Microscope (SEM)

Electron Gun

It is used to produce the electron Beam. Generally, of 2 types:

- Thermionic Emission Gun
- Field Emission Gun

Thermo electrons emitted from a filament (Cathode) made of a thin tungsten wire, (about 0.1mm) by heating the filament at high temperature (about 2800K).

This is the most common form of an electron gun, where in the work function of the materials overcome by the surface temperature of the filament.

Typically, a hairpin filament made of a tungsten wire of 100microm, with a tip radius of 100micro meter, is used to generate the electron beam.

The filament is at negative potential and heated to about 2000 – 2800 k by resistive heating.

The electrons are confined and focussed by a grid cap (wehnelt) held at a slightly higher negative potential than the filament.

The confined beam is accelerated to the anode held at the ground potential and a portion of the beam is passed through a hole.

The filament characteristics are given in terms of brightness which is given as current/area (solid angle). The brightness is of the order of 10^5 (A/cm^2) for tungsten.

More than tungsten brightness needed, Lanthanum hexaboride (LaB_6) and still large brightness needed, we can use field emission gun.

The field emission gun, the electron beam can be achieved by reducing the emission area, which results in significant reduction in the demagnification required in the operation.

Electron Lense/Condenser Lense

An electromagnetic lens used in electron microscopy. This consists of a coil of wire generating a magnetic field enclosed in an iron casing.

Figure 2.3: Condenser Lens

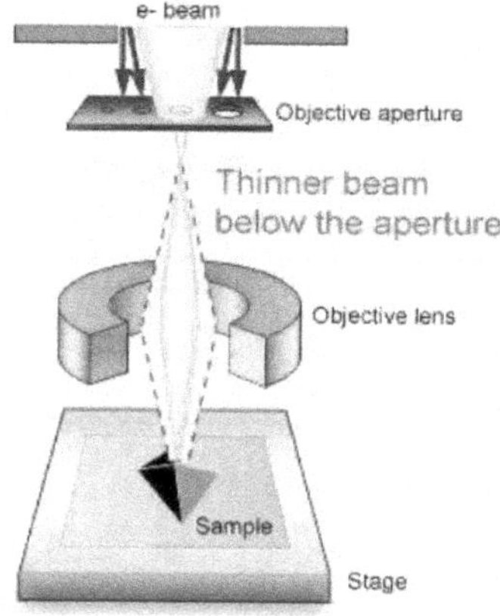

Figure 2.4: Objective Lens

It adjusts the diameter of electron beam made of magnets capable of bending the path of electrons. A magnetic field is generated between the pole pieces by applying a current through a coil. Then the electron in magnetic field undergoes rotation. As the radial component of the magnetic field reverses after the centre of the lens. The rotation in the first half of the lens is reversed.

The electron leaves the lens without any net change in the angular momentum, but it undergoes deflection towards the axis.

An electron micro scope uses 2 kinds of lenses.

- **Condenser lens** → has a large bore giving and long focal length
- **Objective lens** → strong field of short axial extent giving a short focal length resulting in high magnification.

The focal length is given by $f = Kv / (NI)^2$

K = constant

v = Relativistically corrected accelerating voltage

NI = Amp. Turns

Relativistically → moving at a velocity such that there is a significant change in properties (such as mass) in accordance with the theory of relativity a relativistic electron.

Image quality is affected by varying the electron velocity while displaying the image some problems associated with this →

- Lens aberration: can't be solved but reduced to some extent.
- Spherical aberration: image at corners blurs, rather than cutter.
- Chromatic aberration: can be reduced by stabilizing the energy of the electron beam.
- Aperture diffraction: when light wave travel from a narrow source, it does not produce a bright light at centre.
- Astigmatism: improper shape of beam, point object is focussed to 2 – line foci.

Scan Coils

In SEM, the scanned image is focussed by point by point and the scan is achieved by scan coil.

Scan coil consists of 2 solenoids oriented in such a way as to create 2 magnetic fields perpendicular to each other. (X and Y axis). The scan coils lie with in the column and move the electron beam as per requirement across the specimen.

The scan generator changes the step current to the scan coil. The current is then multiplied by a constant by the magnification module and sent to coils.

Varying the current in on solenoid causes the electrons to move left to right.

Varying the current in other solenoid forces these electrons to move from right to left.

Chamber

The place at which specimen will be examined.

Figure 2.5: Chamber

- Structure is very sturdy and insulted from vibration.

- Adjusts the angle and direction of specimen to avoid re-mounting.

- Rugged construction.

Detectors

The detectors are called 'Eyes' of SEM.

These detect the various ways that the electron beam interacts with sample object.

2 kinds of electrons coming out from the sample in SEM.

- Back scattered electrons, with high energies being large or lower than primary beam energy. And

- The secondary electrons which have energies of a few ev or less than few tens of ev.

Most used detectors inside the SEM are secondary electron detector. i.e Everhart – Thorley (E-T) detector.

Depending upon the accelerating voltage and sample density, the signals come from different penetration depths.

In general practice, the detector, a thin metal coating is applied on the surface of scintillator. A high +ve potential is applied to metal surface, so that all electrons including low energy secondary electrons are accelerated to it so as to generate photons. The high voltage should not effect the primary beam of electrons and for this reason, a Faraday cage is kept over the scintillator, by applying a voltage of -50v to +250v, to faraday cage, a complete rejection or collection of secondary electrons becomes possible.

Rutherford Elastic Scattering: When an electron from the beam encounters a nucleus in the specimen, the resultant attraction produces a deflection in the electron's path. Scattering angle is strongly dependent on the atomic number of nuclei involved.

Image Display Unit

Nothing but output of SEM. Scanned image can be displayed in this.

- Fast speed is used for observation of scanned image or specimen
- Slow scan speed is used for saving of image
- Image recorded in digital format
- Nowadays, computers are used where as CRT's were used previously.

Sample Characteristics Detected by a SEM

Topography: The surface features of an object or "how it looks", detectable features are limited to a few mano meters.

Morphology: The shape, size and arrangement of particles making up the object that are lying on the surface of sample or have been exposed by grinding or chemical etching.

Composition: The element and compounds the sample is composed of and their relative ratios, in areas ~ 1nm in dia and depth.

Crystallographic Information: The arrangement of atoms in the specimen and their degree of order. Only useful on single – crystal particles > 20 micrometres.

Sample Preparation

- Cleaning the surface of the specimen.
- Stabilizing the specimen.
- Rinsing the specimen.
- Dehydrating the specimen.
- Drying the specimen.
- Mounting the specimen.
- Coating the specimen.

Applications of SEM

- Detection and analysis of surface fractures
- Examination of surface contaminations

- Providing information in microstructures
- Revelation of spatial variations in chemical compositions
- Identification of crystalline structures
- Wide usage in life science, biology, gemology, medical, forensic science and metallurgy
- Semiconductor inspection
- Assembly of microchips

Advantages of Scanning Electron Microscope (SEM)

- Wide array of applications
- Gathering of versatile information
- User friendly operation
- Generation of date in digital form
- Minimal preparation actions
- Ultra-speed capacity.

Disadvantages

- Expensive, very large in size.
- Sensitivity towards electric, magnetic or vibration interference.
- Requirement of special training personnel.
- Limitations of solid / inorganic samples.

Magnification of Image

$$\text{Image magnification} \quad M \quad = \quad \frac{Scan\ length\ on\ CRT}{scan\ length\ on\ specimen}$$

$$\text{Maximum Magnification} \quad M \quad = \quad \frac{Pixel\ size\ on\ CRT}{Beam\ diameter}$$

Magnification in a SEM can be controlled over a range of about 6 orders of magnitude from about 10 to 3,000,000 times. Unlike optical and transmission electron microscope image magnification in a SEM is not a function of the power of objective lens.

2.3.2. *Transmission Electron Microscope (TEM)*

Transmission electron microscopy can provide micro structure, crystal structure as well as micro-chemical information with high spatial resolution from each of microscopic phases individually.

TEM is powerful tool for materials characterization. To observe the image of electrons emitted from a cold cathode on a florescent screen, Ruska and knoll succeeded in magnifying the image by 17x by using a second solenoid coil between the first lens and the final plate. This can be considered as first prototype TEM to demonstrate the focussing effect of transmitted electrons to form magnified images.

TEM is having capability of recording images of over a million times in magnification with field emission gun electron sources and by using aberration corrected lens, resolution less than 0.1 nm have been achieved.

Magnifications of 400 000 times can be easily obtained for many materials and atoms can be seen at magnifications greater than 15 million times. Materials for TEM must be specially prepared to thicknesses that will allow electrons to be transmitted through the sample, much like light is transmitted through materials in light microscopes.

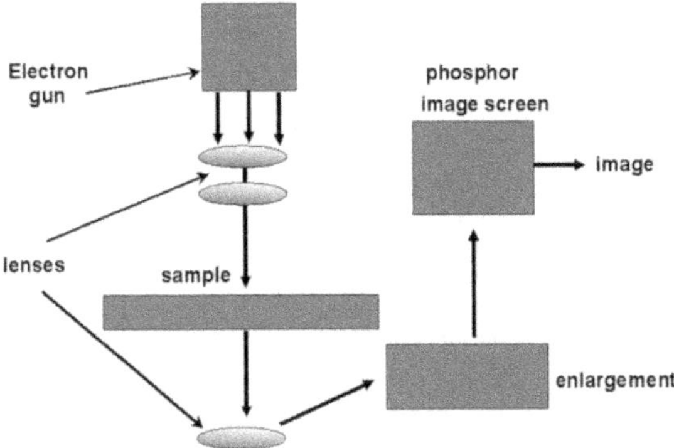

Figure 2.6: Block Diagram for Modern Transmission Electron Microscope

Components of TEM

- Electron Gun
- Condenser Lenses
- Objective Aperture
- Objective Lenses
- Chamber
- Enlargement
- Image Display Unit

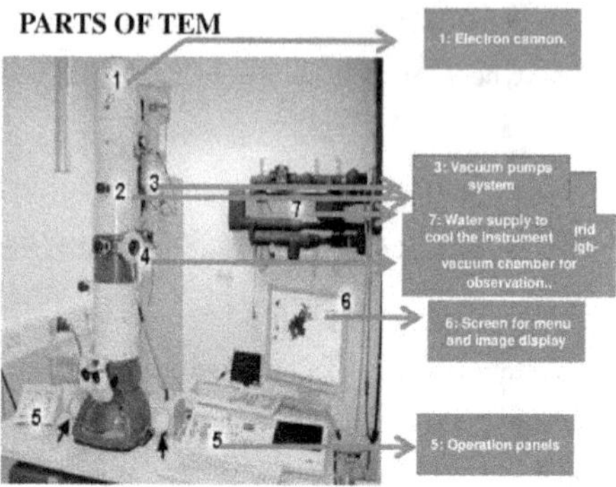

Figure 2.7: Parts of TEM

Electron Gun

Produces electron beam, generally of two types:

A)   Thermionic Emission Gun

B)   Field Emission Gun

Thermo electrons emitted from a filament (cathode) made of a thin tungsten wire, (about 0.1 mm) by heating the filament at high temperature (about 2800k)

The Speed and direction of electrons are controlled through a hole and electrode.

Condenser Lenses

There are Two lenses.

The first lens largely determines the 'spot size', which is the general size range of the final spot that strikes the sample. The second lens changes the size of the spot on the sample, altering it from a wide dispersed spot to a pinpoint beam.

Objective Aperture

It enables the user to examine ordered arrangements of atoms in the sample.

Objective Lenses

Transmitted portion of electrons are focused by the objective lens into an image.

Chamber

- Specimens to examine are placed here
- Structure is very sturdy and insulated from Vibration

- Adjusts the angle and direction of specimen to avoid re-mounting
- The electron beam strikes the specimen and parts of it are transmitted.

Enlargement

The image is passed through lenses and enlarged.

Image Display Unit

The Enlarged image strikes the phosphor image screen and light is generated, allowing the user to see the image. The darker areas of the image represent the areas of the sample through which fewer electrons were transmitted (they are thicker or denser). The lighter areas of the image represent those areas of the sample through which more electrons were transmitted (they are thinner or less dense).

Sample Preparation for TEM

1. Cleaning the surface of the specimen.
2. Stabilizing the specimen.
3. Rinsing the specimen.
4. Dehydrating the specimen.
5. Drying the specimen.
6. Mounting the specimen.
7. Coating the specimen.

Advantages of TEM

- Offer very powerful magnification and resolution.
- Have wide range of applications and can be utilised in a variety of different scientific, educational, and industrial fields.
- TEM's provide information on element and compound structure.
- Images are high quality and detailed.

Disadvantages of TEM

- Very large in space and very expensive
- Laborious sample preparation
- Operation and analysis require special training
- Samples are limited to those that are electron transparent
- TEMs required special housing and maintenance.

- Images are black and white only.

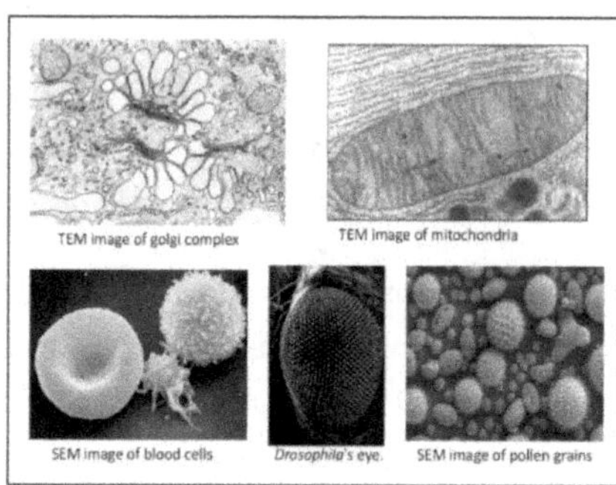

Figure 2.8: Scanned Images of SEM & TEM

2.4.   Scanning Probe Microscopy (SPM)

Scanning probe microscopy is a collective term used for Atomic Force Microscopy (AFM) and Scanning Tunnelling Microscopy (STM).

They have a common operation.

2.4.1. Atomic Force Microscopy (AFM)

The atomic force microscope (AFM) or Scanning force microscope (SFM) is a very high-resolution type of scanning probe microscope, with demonstrated resolution of fractions of a nano-meter, more than 1000 times better than optical diffraction limit.

The AFM and prior STM, were developed by Gerd Binning and Henrich Rohrer in 1981 for which they received Nobel Prize at 1986. The AFM is one of the foremost tools for imaging, measuring and manipulating matter at the nanoscale. The term 'microscope' in the name is actually a misnomer (a wrong or inaccurate name) because it implies looking, while in fact the information is gathered by feeling the surface with a mechanical probe.

Piezo electric elements that facilitate tiny but accurate movements on electronic commands are what facilitate the very precise scanning.

The family of SPM uses NO lenses, but rather a probe that interact with the sample surface.

- It has simple design with low cost
- Easy to handle
- Automatically resolves images.

Basic Principle of AFM

The AFM works based on the principle of Inter Atomic Forces (Attractive and repulsive forces at atomic level). The AFM consists of a microscale cantilever with a sharp tip (Probe) at its which is used to scan the specimen surface.

The cantilever is made up of Silicon or silicon nitride with a tip radius of nano meters. When the tip brought into sample's proximity area, forces between the tip and sample lead to a deflection of cantilever according to *Hooke's law** **(Law of Physics).**

Then the attractive or repulsive forces of atoms in specimen will be measured and include mech constant forces, Vander Waals forces, capillary forces, chemical bonding, electro static forces, magnetic forces, Casimir forces, and solvation forces etc.

*__Hook's law__ *states that, the force needed to extend or compress a spring by some distance scales linearly with respect to that distance.*

__F = Kx__   where

K = constant factor characteristics of spring

X = small is compared to total possible deformation of the spring

The deflection is measured using a laser spot reflected from the top of cantilever into an array of photodiodes.

The **cantilevers#** are fabricated with piezo resistive elements that acts as a strain gauge. Using a Wheatstone bridge, strain in AFM cantilever due to deflection can be measured, but this method is not as sensitive as laser deflection or interferometry.

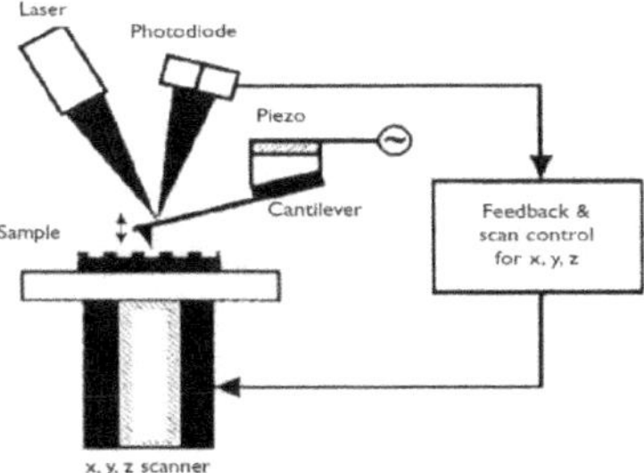

Figure 2.9: Block Diagram of Atomic Force Microscope

If the tip were scanned at a constant height, there would be a risk that the tip would collide with the surface, causing damage. Hence, a feedback mechanism is employed to adjust the tip-to-sample distance to maintain a constant force between the tip and sample.

__Hook's law__ states that, the force needed to extend or compress a spring by some distance scales linearly with respect to that distance.

$$F = Kx$$ where K = constant factor characteristics of spring

X = small is compared to total possible deformation of the spring

#Cantilever: Rigid structural element that extends horizontally and supported at only one end

The sample is mounted on a piezo electrical tube that can move the sample in the Z direction for maintaining the constant force, and X, Y directions for scanning the sample.

A TRIPOD configuration of three piezo crystals may be employed, with each responsible for scanning in the X, Y, and Z directions. Which eliminates some distortions effects seen with tube scanners?

The resulting map of area S = f(X, Y) represents the topography of sample.

Maintaining the distance between cantilever and sample surface is important. Because of, as the cantilever approaches the surface there are weak attractive Vander Waals forces between surface and cantilever atoms. As the cantilever pushed into the surface those become repulsive forces and cantilever may displace the atoms on the surface.

Operating Modes of AFM

The AFM can be operated in a number of modes, depending on application. In general, there are 2 modes.

- Imaging Mode
  - Solid (Contact) Mode
  - Dynamic (Non – Contact) Mode
- Tapping Mode

Possible imaging mode further divided into Static (Contact) Mode and Dynamic (Non-contact) Mode.

Imaging Mode

Contact Mode: In this, the static tip deflection is used as a feedback signal. Because the measurements of a static signal are prone to noise, drift, low stiffness are used to boost the deflection level. When tip is close to surface of sample, attractive forces can be strong, causing the tip to "Snap-in" to surface. Thus, the static mode AFM is almost done in contact mode where the overall force is repulsive.

The force between tip and sample is kept constant during scanning by maintaining a constant deflection.

- The resolution is high.
- Due to short distance between tip & sample, both have chance to get damaged.

Dynamic Mode (Non – Contact Mode)

In this, the cantilever is externally oscillated at or close its resonance frequency. The oscillation amplitude, phase and resonance frequency are modified by Tip-Sample interaction surface. These changes in oscillation with respect to external reference oscillation provide information about the sample's characteristics.

- The distance between the tip and the sample is large and thus the cantilever is attracted. Resolution is low & No damage for tip and sample.

Tapping Mode

In this mode, cantilever is driven to oscillate up and down at near its resonance frequency by a small piezo electric element mounted in AFM tip holder. The amplitude of this oscillation is greater than 10nm, typically 100 to 200nm. Due to the interaction of forces acting on cantilever when tip comes close to surface, Vander Waals force cause the amplitude of this oscillation to decrease as tip gets closer to sample.

- In this mode, the distance between the tip and sample is kept intermediate and thus the cantilever is kept oscillating. The resolution is Better, when compared to non-contact mode.

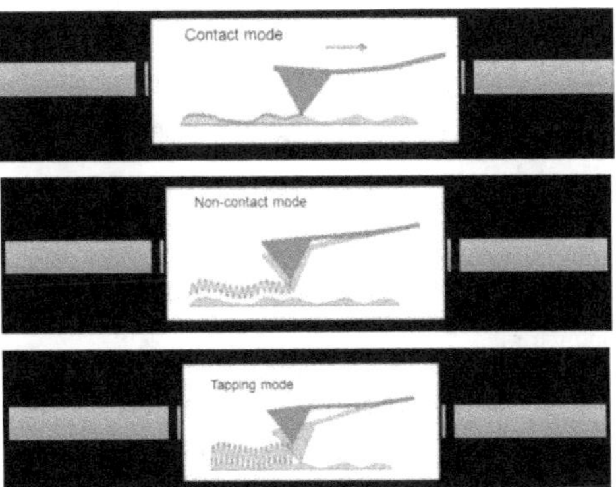

Figure 2.10: Contact, Non-contact and Tapping Modes

2.4.2. *Scanning Tunneling Microscope*

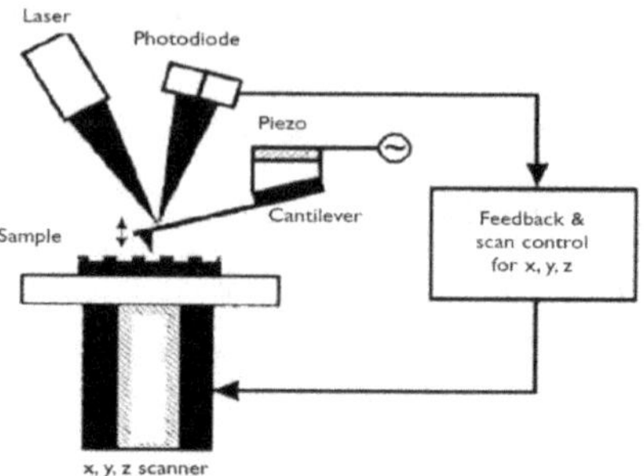

Figure 2.11: Block Dia of Scanning Tunneling Microscope

Components of STM

- **Cantilever:** It has a pointed probe attached to a rectangular base called cantilever. A cantilever is supported at only one end
- **Probe** interacts with the sample surface
- Positioning of cantilever with respect to specimen is achieved by **piezo element**
- **Laser beam** is focussed onto the cantilever
- **A photodiode** is a semiconductor p-n junction device that converts light into an electrical current. In STM, it converts the reflected laser beam into electrical signal. When probe interacts with the sample, cantilever bends and laser reflection from the cantilever changes, resulting in change in the laser spot on the photodiode

STMs are similar to AFMs except they use a sharpened, conducting tip with a bias voltage applied between the tip and the sample. When the tip is brought within about 1 nm of the sample, electrons from the sample begin to pass through the 1 nm gap into the tip or vice versa, depending upon the sign of the bias voltage.

The resulting tunnelling current varies with tip-to-sample spacing, and it is the signal used to create an STM image. For tunnelling to take place, both the sample and the tip must be conductors or semiconductors.

In constant-height mode, the tip travels in a horizontal plane above the sample and the tunnelling current varies depending on topography and the local surface electronic properties of the sample. Similarly, the tunnelling current measured at each location on the sample surface constitutes the topographic image.

In constant-current mode, STMs use feedback to keep the tunnelling current constant by adjusting the chamber height, which is obtained by changing the height of xyz scanner

The resulting tunnelling current varies with tip-to-sample spacing, and it is the signal used to create an STM image. For tunnelling to take place, both the sample and the tip must be conductors or semiconductors.

2.5. Nanomanipulation

The process of moving atoms or altering atoms is known as nano manipulation.

Nanomanipulation is employed in order to produce extremely precise structures and can be broken into two subcategories:

- Nanofabrication and
- Self-assembly.

The first real example of nanomanipulation was demonstrated by Eigler and Schweizer (1990) who manipulated individual xenon atoms using a STM To accomplish pick-and-place manipulation and assembly, a reliable method of gripping is required.

Gripping at the nanoscale is a challenge because it is difficult to control the balance of forces between the object, the surface, and the tool (gripper) where van der Waals and electrostatic forces may dominate.

The scanning tunnelling electron microscope has a computer-controlled probe that skates across the surface The probe can also be programmed to push against the surface like a finger.

When pushing, the charge separating the tip and its fixed mount creates an electrical current that is proportional to the pressure exerted on the tip. By transmitting this current to the proper computer interface a human can actually do the touching. This instrument is called a nanomanipulator. When the scientist gets to a bump on the surface, he or she feels it. In this way scientists can feel surfaces. They have even felt viruses.

A multipurpose nanomanipulator has been built to perform nanomanipulation studies under vacuum inside a scanning electron microscope as well. It can also be used to build simple objects and may eventually allow the operator to 'feel' when they pick up and move atoms. If this was virtually converted to fingertips, we could indeed build structures as if we build things in our everyday environment. In other words, we could use atoms as if they were really small objects for building, like Lego bricks.

Applications for Nanomanipulation

- carbon nanotubes
- Semiconductor preparation and etching for microchip production
- Microbiological system preparation
- Transmission electron microscopy probe preparation
- Construction of components for molecular and quantum computers

2.6. Nano Tweezers

The Danish research group (Nano hand) has developed nano tweezers, which can be used for both imaging and manipulation of nanosized objects to make cell surgery feasible soon.

These nano tweezer probes consist of two wires tapered consecutively through a nanopipette and kept electrically isolated.

The tweezer structure can be closed with an applied electrical field like a pair of chopsticks to produce a device that grasps and moves molecules or atoms.

Nanotubes are specific carbon structures that grow to create hollow tubes only one or two nano meters in diameter. The tubes are mechanically robust and are also good conductors of electricity, making them ideal for scanning probe techniques.

2.7. Atom Manipulation

Single atom manipulation can be achieved with appropriate selection of the charge (polarity), the magnitude and duration of a voltage pulse applied between the STM tip and the sample surface. By placing a tungsten tip above silicon atoms and applying voltages of 5.5 V to the surface for 30 ms, silicon atoms can be lifted from the surface.

Hydrogen can be removed from hydrogen silicon bonds by scanning a tip over the surface while applying a continuous potential difference of several volts, or by applying rapid pulses. The mechanism of extraction is subtle

2.8. Self-Assembly

Self-assembly is a method of constructing nanoscale device and machines.

The technique works by simulating the way biological systems build molecules, viruses, cells, plants, and animals. Rather than having to shift atoms one by one it would be better if we could get them to line up by themselves.

Self-assembly is the process whereby molecules line up where you want them to go naturally, Practically, it is difficult to control atoms and molecules

Molecules and atoms will go to a site of lowest free energy, which is determined by bond strength and entropy factors.

Molecules will line up on a surface if the surface is smooth and they will also align on a surface even though this involves a decrease in entropy if there are sufficient energy savings involved. They are more likely to form monolayers than multilayers because the energy saving, ΔH, involved in forming the second layer, is less.

One of the most powerful ways of attracting molecules so that their free energy is decreased is charge. The electrical attraction between positively charged organic ions and negatively charged metal ions (or the reverse) is therefore a useful assembly mechanism. The other is hydrogen bonding.

Chemists have known about self-assembly for a long time because they have known about adsorption and about catalysis.

Adsorption: process by which a solid holds molecule of a gas or liquid

Catalysis: Catalysis is the mechanism by which the addition of another material (the catalyst) speeds up the existing reaction.

Self-assembly can be done by

- Assembly by electrostatic forces
- Assembly by Vander Waals
- Assembly by hydrophobic interactions
- Assembly by hydrogen bonds

2.9.   X – ray Diffraction (XRD)

X-ray diffraction analysis (XRD) is a technique used to determine the crystallographic structure of a material. XRD works by irradiating a material with incident X-rays and then measuring the intensities and scattering angles of the X-rays that leave the material.

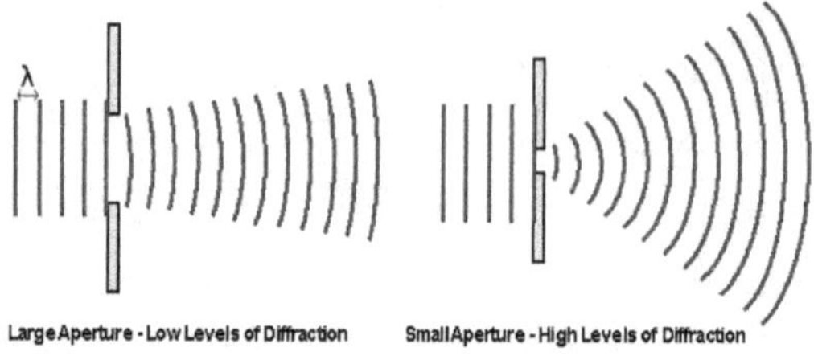

Figure 2.12: Large and Small Aperture

By measuring the angles and intensities of these diffracted beams, a crystallographer can produce a three-dimensional picture of the density of electrons within the crystal. From this electron density, the mean positions of the atoms in the crystal can be determined, as well as their chemical bonds, their crystallographic disorder, and various other information.

Necessity of X-Ray Diffraction

- Can measure the average spacing's between layers or rows of atoms
- Can determine the orientation of a single crystal or grain
- Can find the crystal structure of an unknown material
- Can measure the size, shape, and internal stress of small crystalline regions

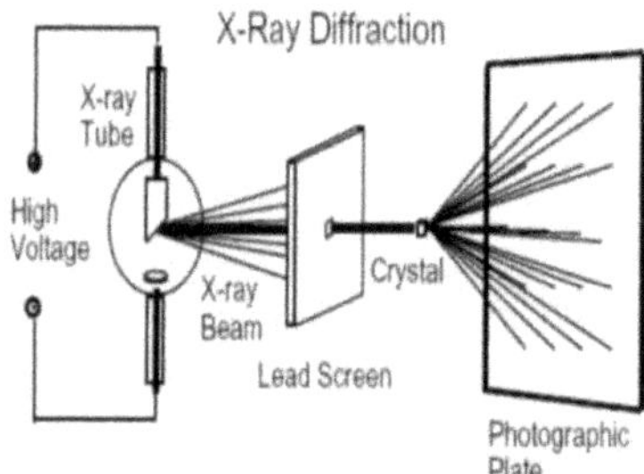

Figure 2.13: X Ray Diffraction

Generally, a typical x-ray diffraction contains below parts:

- Detector
- X-ray source
- Crystal on the end of mounting needle
- Liquid nitrogen steam to keep crystal cold
- Movable mount to rotate crystal

First Step

- In first-step, Researchers crystallize an atom or molecule, because the precise position of each atom in a molecule can only be determined if the molecule is crystallized. If the molecule or atom is not in a crystallized form, the X-rays will diffract unpredictably and the data retrieved will be too difficult.
- The crystal should be sufficiently large (typically larger than 0.1 mm in all dimensions), pure in composition and regular in structure, with no significant internal mistakes such as cracks or twinning.

Second Step

- The crystal is placed in an intense beam of X-rays, usually of a single wavelength (monochromatic X-rays), producing the regular pattern of reflections. As the crystal is gradually rotated, previous reflections disappear and new ones appear; the intensity of every spot is recorded at every orientation of the crystal.

Third & Final Step

- In the third step, these data are combined computationally with complementary chemical information to produce and refine a model of the arrangement of atoms within the crystal. The final, refined model of the atomic arrangement now called a crystal structure is usually stored in a public database.

Advantages

- X-ray is the cheapest, the most convenient and widely used method.
- X-rays are not absorbed very much by air, so the specimen need not be in an evacuated chamber.

Disadvantages

- They do not interact very strongly with lighter elements.
- It is relatively low in sensitivity.

2.10. Spectroscopy

The study of molecular structure and dynamics through the absorption, emission and scattering of light is spectroscopy.

2.10.1. The Raman Spectroscopy Principle

When light interacts with molecules in a gas, liquid, or solid, the vast majority of the photons are dispersed or scattered at the same energy (frequency) as the incident photons. This is described as elastic scattering, or Rayleigh scattering. A small number of these photons, approximately 1 photon in 10 million will scatter at a different frequency than the incident photon. This process is called inelastic scattering, or the Raman effect, named after Sir C.V. Raman who discovered this and was awarded the 1930 Nobel Prize in Physics for his work.

A Raman system typically consists of four major components:

- Excitation source (Laser).
- Sample illumination system and light collection optics.
- Wavelength selector (Filter or Spectrometer).
- Detector (Photodiode array, CCD or PMT).

Unit - III

Nano Powders and Nanomaterials

All materials are composed of grains, which in turn comprise many atoms. These grains can be visible or invisible to the naked eye, depending on their size. Conventional materials have grains varying in size anywhere from hundreds of microns to centimeters. Nanomaterials, sometimes called Nano powders when not compressed, have grain sizes in the order of 1–100 nm in at least one three coordinates typical nanomaterials (zinc oxide and cerium oxide) are shown in Figure 3.1 under transmission electron microscopy.

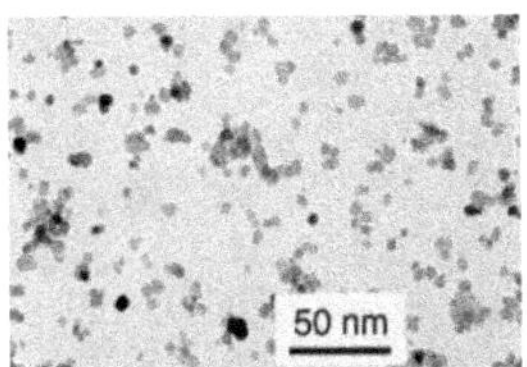

Figure 3.1: Zinc Oxide and Cerium Oxide Under Transmission Micro Scope

Nanomaterials are not new. However, the understanding that certain preparations of oxides, metals, ceramics (a ceramic is defined as an inorganic substance that can be heated into a useful hard structure) and other substances are nanomaterials. Carbon black is a nanomaterial that is used in car tyres to increase the life of the tyre and provide the black colour.

In typical nanomaterials, most of the atoms are located on the surface of the particles, whereas they are in the bulk of conventional materials. Thus, the intrinsic properties of nanomaterials are different from conventional materials, since most atoms are in a different environment. Nanomaterials represent almost the ultimate in increasing surface area. Substances with high surface areas have enhanced chemical, mechanical, optical, and magnetic properties, and this can be exploited for a variety of structural and non-structural applications. In aerospace and automotive applications, for example, materials made from metal and oxides of silicon and germanium exhibit superplastic behaviour, undergoing elongations from 100 to 1000 per cent before failure. This is because the individual nanosized particles can expand relative to each other?

Some nanomaterials are exceptionally strong, hard, and ductile at high temperatures. However, they are chemically very active because the number of surface molecules or atoms is very large compared with the molecules or atoms in the bulk of the material.

Sometimes, to retain the desired properties of the nanomaterial, a stabilizer must be used to prevent further reaction. This enables them to be wear-resistant, erosion-resistant, and corrosion-resistant, but this resistance is usually imparted by some sort of protection mechanism.

3.1. Methods to Produce Nano Materials

There are six widely known methods to produce nanomaterials. These are

- Plasma arcing
- Chemical vapour deposition
- Electrodeposition
- Solgel synthesis
- Ball milling and
- The use of natural nanoparticles.

In plasma arcing and chemical vapour deposition methods, molecules and atoms are separated by vaporization and then allowed to deposit in a carefully controlled and orderly manner to form nanoparticles.

The third method, electrodeposition, involves a similar process, and individual species are deposited from solution. The fourth process, sol-gel synthesis, involves some prior ordering before deposition.

In ball milling, known macrocrystalline structures are broken down into nanocrystalline structures, but the original integrity of the material is retained.

3.1.1. Plasma Arcing

Plasma is an ionized gas. A plasma is achieved by making a gas conduct electricity by providing a potential difference across the two electrodes so that the gas yields up its electrons and thus ionizes. A typical plasma arcing device consists of two electrodes. An arc passes from one electrode to the other. The first electrode (anode) vaporizes as electrons are taken from it by the potential difference.

These positively charged ions pass to the other electrode, pick up electrons and are deposited to form nanotubes. An interesting observation is to make the electrodes from a mixture of conducting and non-conducting materials.

During heating, the non-conducting material is vaporized and ionized so that it also becomes part of the plasma arc and is transported and deposited on the cathode.

To make carbon nanotubes, carbon electrodes are used. Atomic carbon cations are produced. These positively charged ions pass to the other electrode, pick up electrons and are deposited to form nanotubes.

Conditions

- Average mass temperature of arc is higher than 104K, while its power density transferred onto the electrode (anode) is approximately 2KW/mm2.
- Using this method individual carbon nanotubes could be achieved in general several hundred microns long.
- Thus, plasma arcing method is bottom-up method.

3.1.2. Chemical Vapour Deposition (CVD)

CVD involves the dissociation and/or chemical reactions of gaseous reactants in an activated (heat, light, plasma) environment, followed by the formation of a stable solid product.

Chemical vapor deposition (CVD) is a process whereby a solid material is deposited from a vapor by a chemical reaction occurring on or in the vicinity of a normally heated substrate surface

By varying the experimental conditions—substrate material, substrate temperature, composition of the reaction gas mixture, total pressure gas flows, etc.—materials with different properties can be grown.

This method involves depositing nanoparticulate material from the gas phase. Material is heated to form a gas and then allowed to deposit as a solid on a surface, usually under vacuum. There may be direct deposition or deposition by chemical reaction to form a new product. This process readily forms Nano powders of oxides and carbides of metals if vapours of carbon or oxygen are present with the metal.

Production of pure metal powders presents a greater challenge, but it has been achieved using microwaves. In this method, microwaves tuned to metal excitation frequencies are used to melt and vaporize the reactants to produce a plasma at temperatures up to 1500°C. Then the plasma enters a reaction column cooled by water, which facilitates the formation of nanometer size particles.

The residence time of reactants in the plasma can be controlled to ensure complete conversion. Metal concentration in the gas phase (proportional to partial pressure) flow rate of metal vapour, and temperature, alter the particle grain sizes and their distribution. Hence, residence time of reactants in the plasma should be controlled.

After cooling to 700°C or another appropriate temperature these nanoparticles are filtered from the exhaust gas flow at high temperature, where they fall into a container or bottle below. Depending on the material, slow introduction of a stabilizer helps reduce their reactivity by forming a thin protective coating.

Chemical vapour deposition can also be used to grow surfaces. The first layer of molecules or atoms deposited may or may not react with the surface. However, these first formed depositional species can act as a template on which material can grow.

The structures of these materials are often aligned, because the way in which atoms and molecules are deposited are influenced by their neighbours. This works best if the host surface is extremely flat.

Types of Chemical Vapour Deposition (CVD)

- Atmospheric Pressure Chemical Vapour Deposition (APCVD)
- Low Pressure Chemical Vapour Deposition (LPCVD)
- Metal Organic Chemical Vapour Deposition (MOCVD)
- Laser Chemical Vapour Deposition (LCVD)
- Plasma Enhanced Chemical Vapour Deposition (PECVD)

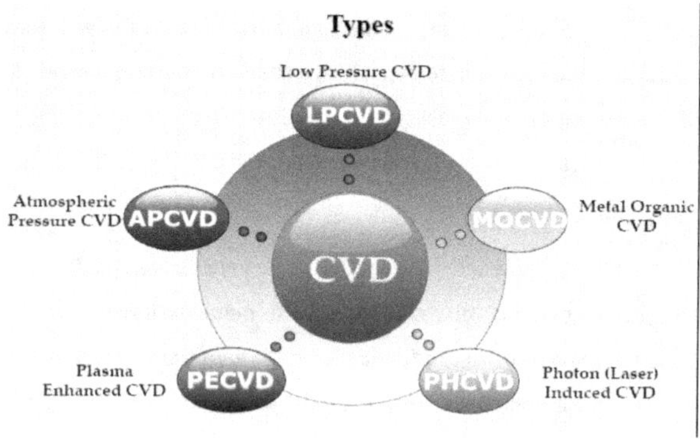

Figure 3.2: Types of CVD

Atmospheric Pressure Chemical Vapour Deposition (APCVD)

- It works at Atmospheric Pressure

- It is used to deposit a layer of material typically several micrometers thick on to a wafer or other type of substrate

- It is also used to/as a surface finishing process for items such as tools and turbine blades, to improve lifetime and performance.

- Since a vacuum system is not required, APCVD system have a relatively low operating cost.

- APCVD has inherently poor utilizations.

- APCVD is extremely susceptible to oxidation due to greater gas density and residence times.

Low Pressure Chemical Vapour Deposition (LPCVD)

- LPCVD coatings exhibit excellent uniformity, high purity and good step coverage.

- It works at sub atmospheric pressures. Reduced pressures tend to reduce unwanted gas phase reactions and improve film uniformly across the wafer.

- The lower pressure increases the precursor diffusion through the gas and the mass transfer rate of the gaseous reactions become higher than the surface reaction rate.

- The pressure for LPCVD is usually around 10-1000 pa. while standard atmospheric pressures is 101,325 pa.

Metal Organic Chemical Vapour Deposition (MOCVD)

- Metal organic compounds are used as molecular precursors to deposit, a wide variety of thin film materials for new industrial applications.

- The great advantage of MOCVD, Precursors are their high volatility at moderate to low temp. therefore reaction temperatures are lower (750 to 1100K) than conventional CVD.

Laser Chemical Vapour Deposition (LCVD)

- LCVD uses a focused laser beam to heat the substrate.

- It also can locally heat a part of substrate while passing the reactant gas, thereby inducing film deposition by locally driving the CVD reaction at the surface.

- It is used to deposit microscale solid patterns or three-dimensional structures on the surface of a substrate by a localized, single step process.

Plasma Enhance Chemical Vapour Deposition.

- PECVD is used to deposit Sio_2, Si_3N_4, (Si_xN_4), $Si_xO_yN_z$ and amorphous Si films.

- Plasma can be used to decompose a molecule that will nor decompose at a reasonable elevated temperature,
- In plasma CVD substrates that can't tolerate high temperature, such as polymers, can be used, where substrate temperatures range from 100 to 500C

Advantages of CVD

- CVD films are generally quite conformal. i.e., the ability of a film to uniformly coat a topographically complex substrate.
- Versatile – any element or compound can be deposited.
- High purity can be obtained.
- High density – nearly 100 % of theoretical value.
- CVD films are harder than similar materials produced using conventional ceramic fabrication processes.
- Material formation well below the melting point.
- Economical in production, since many parts can be coated at the same time.

Disadvantages of CVD

- Chemical and safety hazards caused using toxic, corrosive, flammable and/or explosive precursor.
- Extra steps must be taken in handling of the precursors and in the treatment of the reactor exhaust.
- Restrictions on the kind of substrate that can be coated.
- High deposition temperature (often greater than $600^{0}C$) are often unsuitable for structures already fabricated on substrates.
- It leads to stresses in films deposited on materials with different thermal expansion coefficients, which can cause mechanical instabilities in the deposited films.

Applications of CVD

- **Coatings**: Coatings for a variety of applications such as wear resistance, corrosion resistance, high temperature, protection.
- Semiconductors and related devices.
- Fibre optics – for telecommunications.
- Used in the microelectronics industry to make films serving as dielectrics, conductors, passivation layers, oxidation barriers, and epitaxial layers.

- Used in fabrication of many other industrial applications.

3.1.3. Sol-Gel

Sol–gel process is a method for producing solid materials from small molecules. Dissolve the component in a liquid in order to bring it back as a solid in a controlled manner.

Figure 3.3: Sol-Gel

The sol-gel process is very long known since the late 1800s. The versatility of the technique has been rediscovered in the early 1970s when glasses where produced without high temperature melting processes.

Sol-gel is a chemical solution process used to make ceramic and glass materials in the form of thin films, fibres, or powders.

A sol is a colloidal (the dispersed phase is so small that gravitational forces do not exist; only Van der Waals forces and surface charges are present) or molecular suspension of solid particles of ions in a solvent.

A gel is a semi-rigid mass that forms when the solvent from the sol begins to evaporate and the particles or ions left behind begin to join together in a continuous network.

Colloid

A colloid is a mixture that has particles ranging between 1 and 1000 nanometers in diameter, yet are still able to remain evenly distributed throughout the solution. These are also known as colloidal dispersions because the substances remain dispersed and do not settle to the bottom of the container.

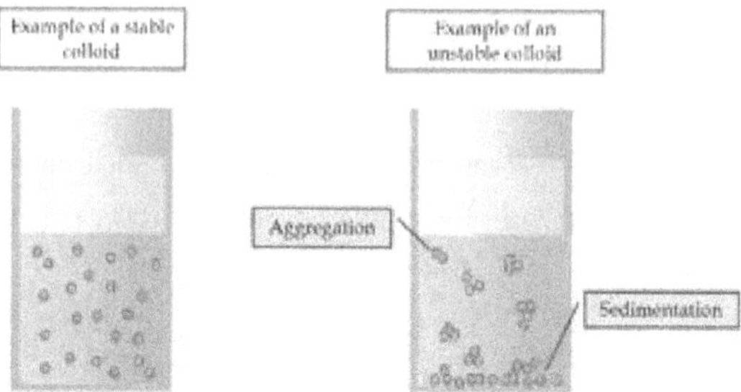

Figure 3.4: Stable Colloid and Unstable Colloid

Sol-gel Formation Occurs in Four Stages

- Hydrolysis;
- Condensation and polymerization of monomers to form particles;
- Growth of particles;
- Agglomeration of particles followed by the formation of networks that extend throughout the liquid medium resulting in thickening, which forms a gel.

Hydrolysis

Chemical breakdown of a compound due to reaction with water. In hydrolysis, the reaction occurs through the addition of water, which results in the replacement of alkoxide groups (OR) with hydroxyl groups (OH).

The Prime Requisite for Obtaining Good Quality in this Technology

- Variation of PH
- Temperature
- Time
- Concentration of Reactants
- Concentration of Catalyst
- Phase transition   Sol to Gel and drying.

Advantages of Sol–Gel

- Can produce thin bond-coating to provide excellent adhesion between the metallic substrate and the top of the sample
- Can produce thick coating to provide corrosion protection performance.
- Can easily shape materials into complex geometries in a gel state.

- Can produce high purity products ceramic oxides can be mixed, dissolved in a specified solvent and hydrolyzed into a sol, and subsequently a gel, the composition can be highly controllable.

- Can have low temperature sintering capability, usually 200-600°C.

- Can provide a simple, economic and effective method to produce high quality coatings.

- Through sol-gel processing, homogeneous, high-purity inorganic oxide glasses can be made at ambient temperatures rather than at the very high temperatures required for conventional formation of glasses

Applications of Sol-Gel

- Protective coatings (Thick and thin)

- Thin films and fibres

- Nano scales powder

- Opto-mechanical process

3.1.4. Electrochemical Deposition

Electrochemical deposition, or electrodeposition refers to a film growth process which consists in the formation of a metallic coating onto a base material (substrate) occurring through the electrochemical reduction of metal ions from an electrolyte to achieve the desired electrical and corrosion resistance, reduce wear and friction, improve heat tolerance, and for decoration.

Electrochemical deposition is typically used to deposit one type of material onto another. For this reason, electrochemical deposition is also sometimes referred as electroplating.

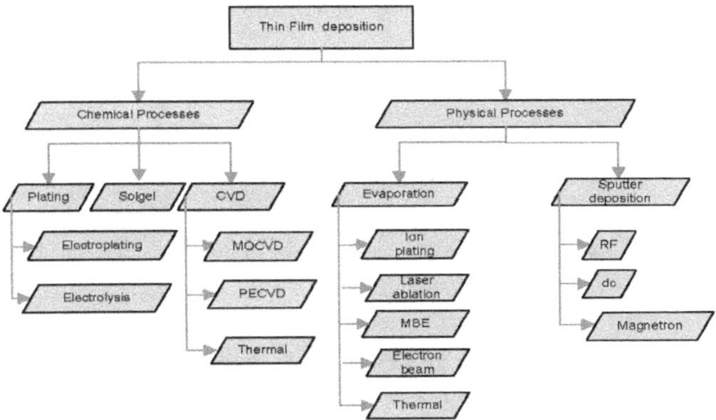

Figure 3.5: Classification of Film Deposition

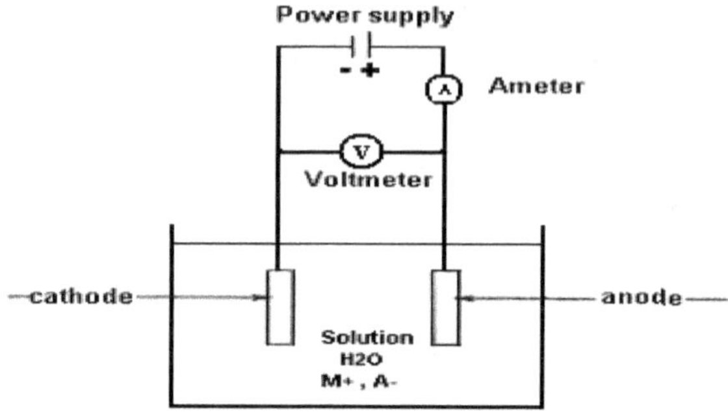

Figure 3.6: Schematic Diagram of Electro Chemical Deposition

- Electrolyte in the electrochemical cell acts as a medium for the movement of electrons and forms the electric circuit between the electrodes
- Oxidation occurs at the anode while reduction reaction occurs at the cathode resulting in electrodeposition
- Oxidation is defined as a process in which an electron is removed from a molecule during a chemical reaction.
- Reduction is defined as a process in which cathode gains electrons
- The flow of ions through the electrolyte onto one of the electrodes causes the electrochemical deposition to occur.
- The electrolyte contains positive and negative ions. Therefore, it is considered an ionic conductor.
- To prepare an electrolyte, the desired metal is mostly dissolved in water. Electrolyte is the material which must deposited

To begin electrodeposition, the cathode (working electrode) immersed in the electrolyte contained in a vessel (cell) along with the anode (counter electrode). To allow electric current flow in the circuit, the two electrodes are connected to a battery or any other power source. The cathode is connected to the negative terminal of the battery, while the anode is connected to the positive terminal so that the metal ions are reduced to metal atoms, which eventually deposit on the surface.

The thickness of the electrodeposited layer on the substrate is determined by the time duration of the plating. The longer the object remains in the chemical bath, the thicker the deposited layer is.

Electrochemical Deposition Conventional Applications

- Aesthetic appeal. Precious metals are expensive. One way to give a cheaper metal the look of a precious metal is to plate it with a thin layer of gold, platinum or silver. This is very common in the jewelry industry.
- To give a material a protective coating. A popular example of this is electroplating steel with a zinc coating. Many different automotive components are manufactured from zinc-plated steel that was coated using the electrochemical deposition process.
- Semiconductor chips are the heart and brain of the Information Age and the Internet Revolution. Everything from supercomputers to personal computers, personal digital assistants, cellular phones, computer games, modern cars, airplanes, household appliances and medical devices need sophisticated chips to run. A common element of these technologies is that they use electrodeposition and other electrochemical micro and nanofabrication methods

Electrochemical Deposition in Nano Technology Applications

Electrodeposition has three main attributes that make it so well suited for Nano technology

- It can be used to grow functional material through complex 3D masks.
- It can be performed near room temperature from water-based electrolytes.
- It can be scaled down to the deposition of a few atoms or up to large dimensions

Electrode Potential

When a metal electrode is placed in an ionic solution, ions will be exchanged between the metal and the solution. Ions from the metal enter the solution, and ions from the solution enter the metal lattice. The boundary between the two phases is called the interface. Due to an electron exchange between the electrode and the ions in solution, the interface becomes electrified and a potential difference across the metal-solution interface is generated. This potential difference is called electrode potential.

3.1.5. Ball Milling

It is efficient tool for grinding many materials into fine powder. A ball mill is a type of grinder used to grind and blend materials for use in mineral dressing processes, paints, pyrotechnics, ceramics, and selective laser sintering etc. A ball mill works on the principle of impact and attrition (churning).

Impact

Size reduction is done by impact as the balls drop from near the top of the shell.

Figure 3.7: Ball Milling Machine

A ball mill consists of a hollow cylindrical shell rotating about its axis. The axis of the shell may be either horizontal or at a small angle to the horizontal. It is partially filled with balls. The grinding media are the balls, which may be made of steel (chrome steel), stainless steel, ceramic, or rubber. The inner surface of the cylindrical shell is usually lined with an abrasion-resistant material such as manganese steel or rubber lining.

- The length of mill is approximately equal to its diameter.
- Ball occupies about 30-50% of the volume.
- The diameter of ball is 12mm to 125mm. The dia of shell is 3m and its length is about 4.25m.
- The shell is rotated through a drive gear with speed 60-100 rpm.

Wet or dry ball mill grinding machine consists of following parts:

- Feeding part
- Discharging part
- Turning and driving part (gear, motor etc)
- Rotating balls

In case of continuously operated ball mill, the material to be made to powder is fed from the left through a 60° cone and the product is discharged through a 30° cone to the right. As the shell rotates, the balls are lifted up on the rising side of the shell and then they cascade down (or drop down on to the feed), from near the top of the shell. In doing so, the solid particles in between the balls and ground are reduced in size by impact.

Applications of Ball Milling

- Generation of curved or closed shell carbon nano structures by ball milling of graphite.
- Carbon scrolls produced by high energy ball milling of graphite.
- Non porous carbon produced by ball milling.
- The nucleation and growth of carbon nano tubes in a mechano-thermal process.
- Carbon nano tubes formed in a graphite after mechanical grinding and thermal annealing.
- Carbon microspheres produced by high energy ball mill.
- Low energy pure shear milling. A method for the preparation of graphite nano-sheets.

Advantages of Ball Milling

- The cost of installation is low.
- The cost of production is low.
- It is suitable for materials of all degree of hardness.
- The grinding medium is cheap.
- Suitable for both batch and continuous operations.

Dis Advantages of Ball Milling

- Rotational ball loading difficulty
- Mill efficiency is low
- The mill feed size range is wide
- Not easy to get thinner product

3.1.6. Natural Nano Materials

- The nano material which belongs to natural world is called as natural nano materials.
- These are having remarkable properties due to inherent nano structures.
- No involvement of human modification and/or processing.

Forming Nano Particles from Natural Sources

- Chemical precipitation (Bottom-up)
- Abiotic (bio-arrested)
- Thermal
- Solar
- Bio mass combustion (T, O_2)

- Cosmic dust

- Volcanic activity

- Weathering actions and wind erosion (Top-down)

Natural nano particles consist of aerosols, colloids, and these natural nano particles are constituents of soil and sediments.

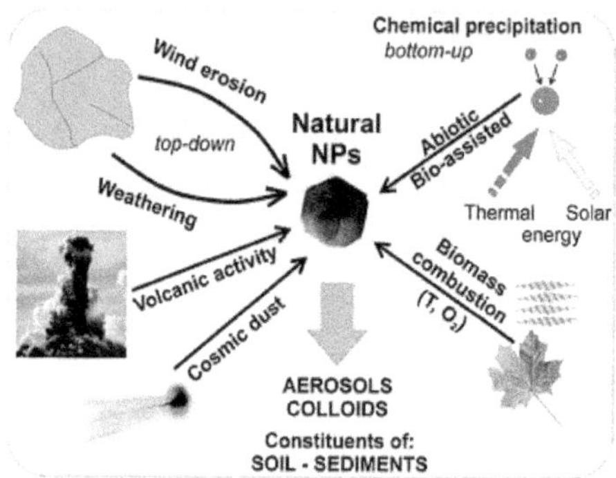

Figure 3.8: Aerosols and Colloids

Bio Mimicry

Bio Mimicry explains the examination of nature, its models, systems, processes, and elements to emulate or take inspiration from in order to solve human problems.

Ex: By observing the self-cleaning of a lotus Leaf, we have created self-cleaning glasses in traffic control units

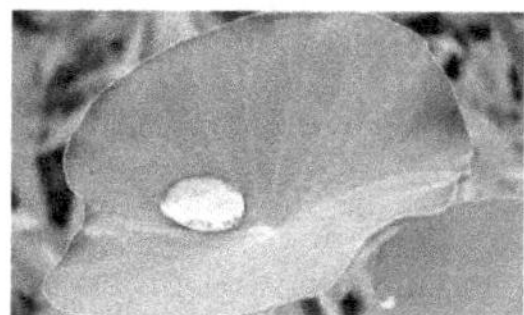

Figure 3.9: Self-cleaning of Lotus Leaf

Gecko's Sticky Feet

- Geckos are lizards belonging to the infraorder gekkotan.

- Can bring to any surface at any orientation.

- Can walk on smooth and rough surfaces.

- Upside down on a glass surface.

- Walk on a dirty or wet surface maintaining full contact.

Water Strider

A large water-repellent force was produced by nano structures on the water striders leg.

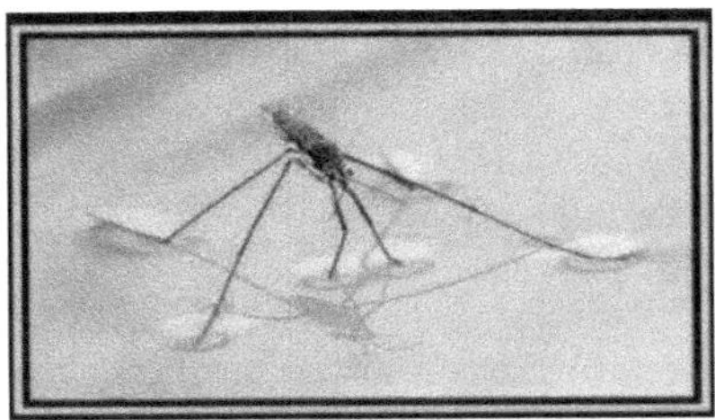

Figure 3.10: Water Strider

Water striders able to walk on water for a number reasons like

- Surface area
- Gravitational forces
- Surface forces (Vander Waals forces)
- A waxy (hydrophobic) surface on their legs.
- Most importantly micro hairs on their feet are 'nano-groovy'

SEM Analysis of Wings

Shows even more intricate structure called SETAE: looks like fir trees. About 400 nm long, responsible for producing constructive interference in blue wavelengths which generates strong blue colour.

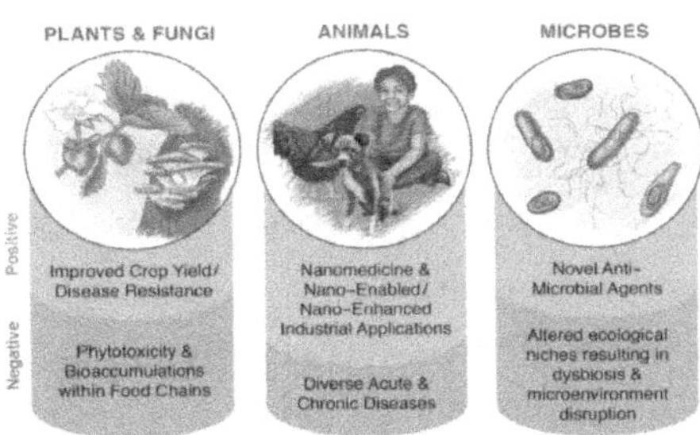

Figure 3.11: Nano Particle Impact

Nano Materials from Natural Nano Particles

There has been considerable interest in producing pores in materials that moderately sized small molecules of 10 nm or so can fit into and react on the surface. Bombarding materials with energetic heavy ions accelerated by a cyclotron is one way of doing things. cyclotron accelerates charged particles in a spiral path, which allows for a much longer acceleration path than a straight-line accelerator

Applications of Nanomaterials

- Nanocrystalline materials synthesised by the sol-gel technique result in aerogels.
- Aerogels are ultra-porous materials, which means that although they are solid, they are full of tiny air-filled holes called pores. Those pores are the key to aerogels' unique properties.

Aerogels

- Since they are porous and air is trapped at the intersect, aerogels are currently being used for insulation in offices, homes, and so on.
- By using aerogels for insulation, heating, and cooling bills are drastically reduced, thereby saving power and reducing the attendant environmental pollution.
- They are also being used as materials for 'smart' windows, which darken when the sun is too bright and lighten when the sun is not shining brightly.

3.2. Machine Tools

Some nanomaterials are harder than conventional materials. Cutting tools and drills made of nanocrystalline materials, such as tungsten carbide, tantalum carbide, and titanium carbide, are much harder, much more wear-resistant, erosion-resistant, and last longer than their conventional (large-grained) counterparts.

They also enable the manufacturer to machine various materials much faster, thereby increasing productivity and significantly reducing manufacturing costs.

Nanocrystalline silicon nitride (Si3N4) and silicon carbide (SiC), have been used in automotive applications as high-strength springs, ball bearings, and valve lifters, because they are readily machinable and have excellent physical, chemical, and mechanical properties. They are also used as components in high temperature furnaces.

On the other hand, some nanomaterials are softer than conventional materials. Conventional ceramics are very hard, brittle, and difficult to machine. Zirconia, ZrO2, a hard, brittle ceramic, has even been rendered superplastic by nanocrystalline grains. It can be deformed to great lengths (up to 300% of its original length).

Nanocrystalline ceramics can be pressed and sintered into various shapes at significantly lower temperatures, whereas it would be very difficult, if not impossible, to press and sinter conventional ceramics even at high temperatures.

3.3. Batteries

Nanocrystalline materials synthesised by sol-gel techniques are used as separator plates in new generation batteries because of their aerogel structure, which can hold considerably more energy than conventional plates.

Nickel metal hydride (Ni-MH) batteries made of nanocrystalline nickel and metal hydrides require far less frequent recharging and last much longer.

3.4. Powerful Magnets

The strength of a magnet increases with increased surface area per unit volume. It has been shown that magnets made of nanoparticles possess very unusual magnetic properties due to their extremely large surface area.

Typical applications for these high-power rare-earth magnets include quieter submarines, automobile alternators, land-based power generators, motors for ships, ultra-sensitive analytical instruments, and magnetic resonance imaging (MRI) in medical diagnostics.

Unit - IV

Nano Electronics

Nano electronics refers to the use of nano technology in electronic components. Nano electronics covers a diverse set of devices and materials with the common characteristics that they are so small that inter-atomic interactions and quantum mechanical properties need to be study.

Nano electronics devices have critical dimensions with a size range between 1nm to 100nm.

For ex. Silicon MOSFET (silicon metal oxide semiconductor field effect transistor) including 22nm CMOS and succeeding 14nm, 10nm, 7nm Fin FET (Fin field effect transistor) generations.

Nano electronics is sometimes considered as '**Disruptive Technology**' because there are different from traditional transistors.

Nanoelectronics is the term used in the field of nanotechnology for electronic components and research on improvements of electronics such as display, size, and power consumption of the device for the practical use.

This includes research on memory chips and surface physical modifications on the electronic devices. Improving display screens on electronics devices. This involves reducing power consumption while decreasing the weight and thickness of the screens. Increasing the density of memory chips.

Researchers are developing a type of memory chip with a projected density of one terabyte of memory per square inch or greater. Reducing the size of transistors used in integrated circuits.

One researcher believes it may be possible to **"put the power of all of today's present computers in the palm of your hand"**.

4.1. History of Nano Electronics

In 1965, Gordon Moore observed that silicon transistors were undergoing size reduction. And it was confined later as **"Moore's Law."** By his observation, transistor minimum feature sizes have been decreased from 10 micro meters to 10 nano meters as of 2019.

The field of nano electronics aims to enable to continued realisation of this las using new methods and materials to build electronics devices with feature sizes on the nano scale.

4.2. Approaches to Nano Electronics

There are 3 approaches to configure nanos electronic devices

- Nano fabrication
- Nano materials electronics
- Molecular electronics

Nano Fabrication

This method is used to develop or design arrays or layers of nano electronic devices to work for single operation.

For ex: electron transistors, which involves transistor operation based on a single electron.

Nano fabrication can be used to construct ultra-dense parallel arrays of nano wires as an alternative to synthesizing nano wires individually. Nano electro mechanical systems (NEMS) fall under this category.

Nano Material Electronics

In this the transistors are packed as arrays on to a single clip. Thus, they remain in a uniform manner and symmetrical in nature. Thus, they are known to have a speedier movement of electronics in materials.

Here, nano particles can be used as quantum dots. Besides being small and allowing more transistors to be packed into a single clip, the uniform and symmetrical structure of nano wires and/or nano tubes allows a higher electron mobility (faster electron movement in the material), a higher dielectric constant (faster frequency), and a symmetrical electron/hole characteristic.

Molecular Electronics

Molecular electronics is a new technology which is still in its starting age, but also brings hope for truly atomic scale electronic system in the future.

Applications of molecular electronics were proposed by IBM researcher Ari Aviran and other, Mark Ratner in 1974 and 1988 respectively.

Other Approaches:

Nano Ionics -- transport of ions rather than electrons in nano scale system

Nano Photonics – studies the behaviour of light on nano scale.

4.3.   Nano Electronic Devices

There are many types of nano electronic devices such as

- Integrated circuits with nano scale range of 50nm.
- Computers
- Memory storage
- Novel opto electronic devices
- Displays
- Quantum computers
- Radios
- Energy production devices.
- Medical diagnostics

Integrated Circuit

Initially, electronic devices such as the transistor radio were made by painstakingly soldering all the individual resistors, capacitors, diodes and transistors onto a circuit board with interconnecting wires.

However, in 1958 Jack Kilby at Texas Instruments realised that if conventional circuit elements, such as resistors and capacitors could be fabricated in silicon, they could be incorporated with transistors onto a single silicon substrate.

As well as miniaturising circuits by consolidating the circuit elements, and doing away with the need for interconnecting wires, this procedure also eliminated assembly (soldering) errors.

Kilby used photographic techniques to transfer patterns to the silicon wafer and introduced precise concentrations of dopants to the semiconductor to specific areas to form a new integrated circuit.

Transistor

The first transistor was about a centimetre high and made of two gold wires separated by 0.02 inches on a germanium crystal. The transistor worked by applying a potential to the 'emitter' wire, which modulated and amplified the current between the 'collector' wire and the base electrode (the germanium crystal).

Bardeen and Brattain won the Nobel Prize in Physics for its invention in 1956. By today's standards the size of the original transistor is mammoth — a modern processor contains transistors that are ten thousand times smaller. We have come a long way in the last 50 years from this single, bulky and unreliable transistor to the more than 42 million transistors in a modern processor, where each and every transistor has to work reliably.

A transistor is a semiconductor device used to amplify or switch electrical signals and power. The transistor is one of the basic building blocks of modern electronics. It is composed of semiconductor material, usually with at least three terminals for connection to an electronic circuit.

4.4.   Tools for Micro and Nano Fabrication

Micro & Nano fabrication will be done by 3 methods.

- Optical lithography or Photo Lithography
- Electron beam lithography
- Atomic lithography

4.4.1. *Optical Lithography or Photo Lithography*

It is a method of printing originally based on immiscibility of oil and water.

Immiscibility means that incapable of being mixed or blended. Even when we mixed both, it cannot from a homogeneous mixture.

Ex: oil, Benzene

Optical or Photo Lithography means which used the usage of photons. Lithography is invented in 1796 by Bavarian Author **Johann Alios Senefelder** (1771-1834) as a cheap method of publishing work. Lithography can be used to print text or artwork onto a paper or other suitable substrate or material.

For lithography processing, a hard copy of the pattern must be first generated. This is called a reticle or mask. The design on the mask must be transferred to the wafer.

First, the pattern is transferred to a photoresist layer on the wafer.

Second, the transfer of the pattern takes place from the photoresist to the wafer. This stage is final and it is very hard to remove the formed patterns without causing damage to the underlying wafer. This process is called developing.

Then, the transfer of the pattern takes place from the photoresist to the wafer. This stage is final and it is very hard to remove the formed patterns without causing damage to the underlying wafer.

Photolithography is widely used in the integrated circuits (ICs) manufacturing. The process of IC manufacturing consists of a series of 10-20 steps or more, called mask layers, where layers of materials coated with resists are patterned then transferred onto the material layer.

Photo Resist

A photoresist is a light-sensitive material used in several processes, such as photolithography and photoengraving, to form a patterned coating on a surface. This process is crucial in the electronic industry in which all integrated circuits, chips will be designed.

The process begins by coating a substrate with a light sensitive organic material. A patterned mask is then applied to the surface to block light, so that only unmasked regions of material will be exposed to light. A solvent called a developer, is then applied to the surface.

A photoresist is alight sensitive polymer, when exposed to UV light, it turns to a soluble material developer those exposed areas can then be dissolved by using a solvent, leaving behind the pattern.

The solubility of a substance, in the max amount of a material called the solute, that can be dissolved in given quantity of specified solvent at a given temperature.

Photo resists are radiation sensitive materials that usually consists of a phot sensitive compound, a polymeric backbone, and a solvent.

Photo resists can be classified upon their solubility after exposure into,

- Positive resist (solubility of exposed area increases)
- Negative resist (solubility of exposed area decreases)

Positive Photo Resist

A positive photoresist is a type of photoresist in which the portion of photoresist that is exposed to light becomes soluble to the photoresist developer. The unexposed portion of the photoresist remains insoluble to photo resist developer.

The resist is exposed with UV light where the underlying material is to be removed. In these resists, exposure to the UV light changes the chemical structure of resist so that it becomes more soluble in developer. The exposed resist is then washed away by the developer solution, leaving windows of the base underlying material. The mask, therefore contains exact copy of pattern which is to remains on the wafer, as a standstill for subsequent processing.

Ex: MCC – PMMA Series (e-beam)

Dupont – S1800 series (g-line)

Dupont – SPR – 220 (i-line)

Negative Photo Resist

A negative photo resist is a type of photo resist in which the portion of the photoresist that is exposed to light becomes insoluble to the photoresist developer.

The unexposed portion of photo resist is dissolved by photo resist developer.

Negative photoresists, exposed to UV light, the negative resist becomes (crosslinked / polymerised) and more difficult to dissolve in developer. Therefore, the negative resist remains on the surface of substrate where it exposed. And developer solution removes only the unexposed areas.

Masks used for negative photoresists, therefore contain the inverse or photographic "negative" of the pattern to be transferred.

Ex: MCC – SU8 Series (i-line)

MCC – KMPR Series (i-line)

Dupont – UVN-30 (DUV)

MRT – ma – N 1400 Series (i-line)

MRT – ma – N 2400 Series (DUV)

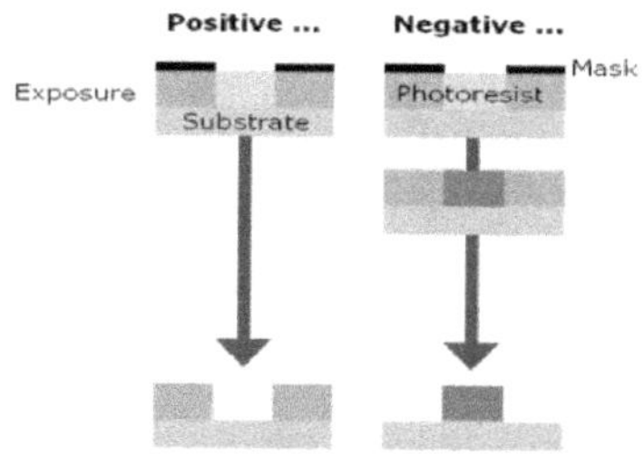

Figure 4.1: Positive and Negative Photo Resist

This lithography process will be used to fabricate semiconductors chips and circuits. In semiconductor lithography, a thin layer of polymer (known as a resist) that is sensitive to ultraviolet radiation.

Using a mask that contains the pattern required for the transistor fabrication, radiation is projected through the mask onto the resist thereby altering the resist's chemical nature and hence solubility.

The ultraviolet light alters the resist layer chemically so that the exposed area becomes more soluble (positive resist) or less soluble (negative resist) in a developing fluid.

Using conventional optics, a much-reduced image of the mask is thus projected onto the resist layer, which is then developed away to reveal the clean semiconductor surface in the exposed (positive resist) or unexposed (negative resist) regions.

Overview of an Optical Lithography Process

- A silicon wafer is covered with light sensitive resist;
- Ultraviolet light is projected through a mask;
- The exposed region becomes more soluble in a developing fluid so that the pattern is reproduced on the resist layer;
- The underlying semiconductor material can be etched away in the regions of the holes using an acid;
- A solvent can be used to wash away the resist; and
- The pattern is now transferred to the semiconductor material.

4.4.2. Electron Beam Lithography

Electron Beam Lithography is a higher energy, shorter wave length radiation lithography systems used in research laboratories. By this lithography smaller devices can be fabricated.

The resolution limits in these systems are determined by a variety of factors, including the spatial distribution of the deposited energy. The granularity of the resist employed. The contrast of resist developer and the statistical distribution of photons at each pixel.

Electrons are emitted from the electron gun of a scanning electron microscope and are focused on the sample using electron optics with a resolution of approximately 0.5 nm. However, a different chemical resist is required that is now sensitive to **the wavelength of electron irradiation** rather than ultraviolet light. The resist itself limits the final resolution to approximately 5 nm.

During the lithographic process the whole system is kept under vacuum and a single beam of electrons is focussed at the surface of the resist-coated semiconductor wafer. The electron beam is moved across the surface under computer control using a pattern generator.

Whilst providing higher resolution and greater flexibility than optical lithography, this system has several disadvantages for commercial application. The main problem is the length of time to write the pattern compared to optical lithography, where a whole chip is exposed through the mask.

Another problem is the so-called proximity effect: the high- energy electrons (5–100 kV) undergo a large degree of **elastic scattering** in the resist and substrate so that the region of the electron beam resist that is affected by the electrons is much bigger than the size of the finely focussed electron beam spot. This tends to **blur the pattern**.

4.4.3. *Atomic Lithography*

Atomic lithography uses Scanning Tunneling Microscope. A voltage difference is then applied between the tip and the sample, causing a small number of electrons to tunnel from the tip to the sample surface. The magnitude of this tunnelling current is very sensitive (exponentially) to the separation between the tip and the nearest surface atom.

In order to obtain a topological map of the surface, the tip is carefully moved a small distance across the sample surface. As the tip moves, the tunnelling current changes, reflecting the change in the tip-surface distance.

This change in current is fed back to an electrical circuit that controls the up-and-down motion of the tip. If the tunnelling current decreases, the tip-surface separation must have increased, so the feedback circuit causes the tip to be lowered until the current returns to the previous value. Conversely, if the tunnelling current increases, the tip-surface distance must have decreased so the feedback circuit raises the tip.

In this way the feedback circuit works to keep the tip-surface distance constant as the tip is scanned across the surface. The up and down motion of the tip is then tracked by a computer which creates a topographical map of the height of the tip over the surface, effectively plotting the height of the surface atoms.

STM uses a nanometre-scale tool to move individual atoms around on a surface. Applying a few volts between the tip and surface can result in electric fields that can break local chemical bonds. As a result, a wide variety of local manipulations and modifications can be made ranging from atom displacement, removal and deposition of individual atoms.

Recently it has been shown that it is even possible to control the placement of atoms in semiconductors, where the bonds are extremely strong, to fabricate electronic devices at the nanoscale. After covering a semiconductor surface in an ultra-high vacuum compatible resist, it has been possible to use the tip of the STM to knock off individual atoms from the resist surface, thereby creating a mask of atomic-scale resolution.

4.5.   Molecular Beam Epitaxy (MBE)

Molecular-beam epitaxy (MBE) process was developed in the late 1960s at Bell Telephone Laboratories by J. R. Arthur and Alfred Y. Cho.

Molecular-beam epitaxy (MBE) is defined as growth of epitaxial films on a hot substrate from molecular beams under ultra-high vacuum conditions. Epitaxy refers to a type of crystal growth or material deposition.

MBE is widely used in the manufacture of semiconductor devices, including transistors, and it is considered one of the fundamental tools for the development of nanotechnology.

It is also used for the deposition of some types of organic semiconductors. In this case, molecules, rather than atoms, are evaporated and deposited onto the wafer.

MBE systems can also be modified accordingly to the needs. Oxygen sources, for examples, can be incorporated for depositing oxide materials for advanced electronic, magnetic, and optical applications, as well as for fundamental research.

To make an interesting new crystal using MBE, you start off with a base material called a substrate, which could be a familiar semiconductor material such as silicon, germanium, or gallium arsenide.

First, you heat the substrate, typically to some hundreds of degrees (for example, 500–600°C or about 900–1100°F in the case of gallium arsenide). Then you fire relatively precise beams of atoms or molecules (heated up so they are in gas form) at the substrate from "guns" called effusion cells.

You need one "gun" for each different beam, shooting a different kind of molecule at the substrate, depending on the nature of the crystal you are trying to create. The molecules land on the surface of the substrate, condense, and build up very slowly and systematically in ultra-thin layers, so the complex, single crystal grows one atomic layer at a time. That's why MBE is an example of what's called thin-film deposition.

One reason that MBE is such a precise way of making a crystal is that it happens in highly controlled conditions: extreme cleanliness and what is called an ultra-high vacuum (UHV), so no dirt particles or unwanted gas molecules can interfere with or contaminate the crystal growth. "Extreme cleanliness" means even cleaner than the conditions used in normal semiconductor manufacture; an "ultra-high vacuum".

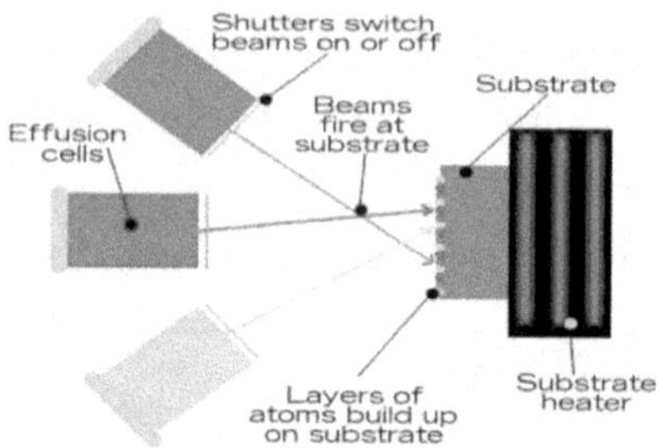

Figure 4.2: Molecular Beam Epitaxy

Advantages of Molecular Beam Epitaxy (MBE)

- To build a solar-cell by depositing a thin film of a photovoltaic material onto a substrate.
- To design a precise thin-film device for computing, optics, or photonics, MBE is one of the best techniques

4.6.   MEMS (Micro-Electro-Mechanical Systems)

Micro Electro Mechanical Systems is the integration of mechanical elements, sensors, actuators, and electronics on a commo silicon substrate through microfabrication technology.

MEMS is a process technology used to create tiny integrated devices or systems that combine mechanical and electrical components.

MEMS devices are consisting 4 parts.

- Micro Electronics
- Micro Sensors
- Micro Actuators
- Micro Structures

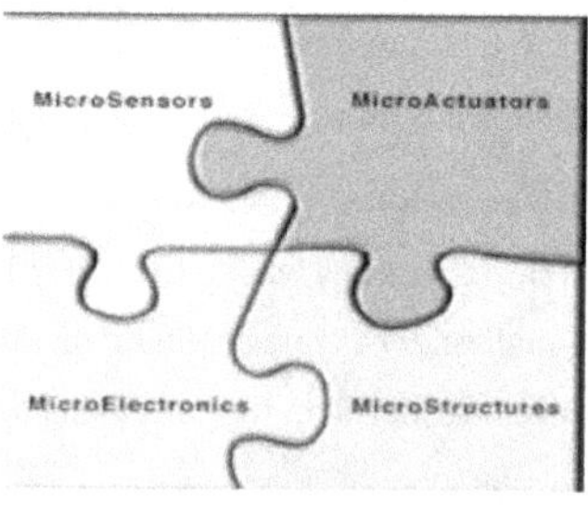

Figure 4.3: Schema of MEMS

78

Micro Electronics

- It is like human brain that receives, processes, and makes decisions.
- Data comes from micro sensors

Micro Sensors

- These will constantly gather data from environment.
- Pass data to microelectronics for processing.
- Can monitor mechanical, thermal, biological, and chemical readings.

Micro Actuators

- Acts as trigger to activate external device.

Microelectronics

- It will tell micro actuator to activate the device.

Micro Structures

- Extremely small structures built into surface of chip.
- Built right into silicon of MEMS.

MICRO SENSORS → MICRO ELECTRONICS → MICRO ACTUATORS

MICRO STRUCTURES

The functional elements of MEMS are miniaturized structures, sensors, actuators, and microelectronics, the most notable elements are the microsensors and micro actuators all integrated onto the same silicon chip. These devices (or systems) can sense, control, and actuate on the micro scale, and generate effects on the macro scale.

Micro Structures Examples:

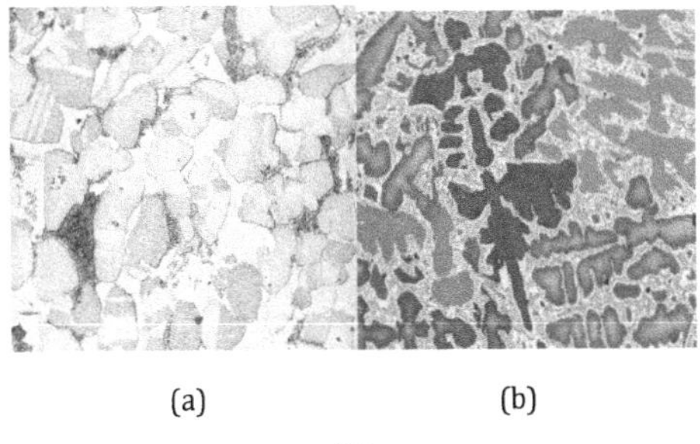

(a)   (b)

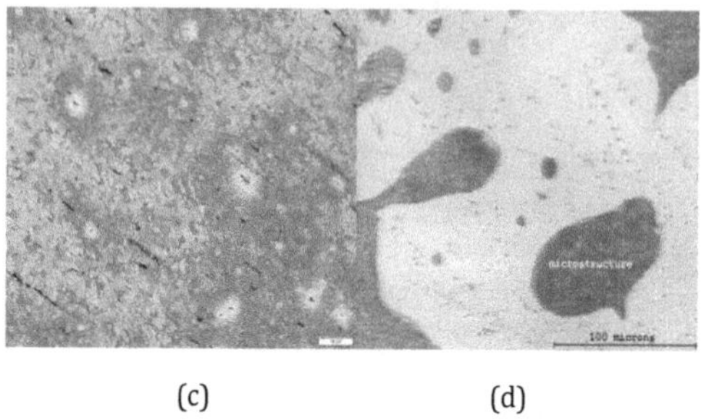

(c)   (d)

Figure 4.4: Micro Structure Examples

(a) Metallography allows the metallurgist to study the microstructure of metals. (b) A micrograph of bronze revealing a cast dendritic structure (c) Al-Si microstructure (d) Micro structure of a spiral tube originating from the 12th-13th century cemetery.

Automotive Airbag Sensor

Automotive airbag sensors were one of the first commercial devices using MEMS. They are in widespread use today in the form of a single chip containing a smart sensor, or accelerometer, which measures the rapid deceleration of a vehicle on hitting an object. The deceleration is sensed by a change in voltage. An electronic control unit subsequently sends a signal to trigger and explosively fill the airbag.

Medical Pressure Sensor

Another example of an extremely successful MEMS application is the miniature disposable pressure sensor used to monitor blood pressure in hospitals. These sensors connect to a patients intravenous (IV) line and monitor the blood pressure through the IV solution.

The MEMS device fabrication will be done by 2 methods

- Bulk Micromachining
- Surface Micromachining

4.6.1. Bulk Micromachining

Bulk Micromachining is a process used to produce micromachinery or micro electro mechanical systems. Bulk micromachining defines structures by selectively etching inside a substrate. This process typically uses wafers of silicon, but will occasionally use plastic or ceramic material as well. Bulk Micromachining starts with a solid piece and removes material until it reaches its final shape, as opposed to surface micromachining, which builds a piece layer by layer. The most common method for performing bulk micromachining is via selective masking and wet chemical solvents. The newer alternative to this method is dry etching using plasma or laser system to remove unwanted material. This is generally more accurate than wet etching, but it is also more expensive.

Steps of MEMs Fabrication Using Bulk Micromachining

- Step1: The first step involves the circuit design and drawing of the circuit either on a paper or on using software like PSpice or Proteus.
- Step 2: The second step involves the simulation of the circuit and modelling using CAD (Computer-Aided Design). CAD is used to design the photolithographic mask which consists of the glass plate coated with chromium pattern.
- Step 3: The third step involves photolithography. In this step, a thin film of insulating material like Silicon Dioxide is coated over the silicon substrate, and over this, an organic layer, sensitive to ultraviolet rays is deposited using spin coating technique. The photolithographic mask is then placed in contact with the organic layer. The whole wafer is then subjected to UV radiation, allowing the pattern mask to be transferred to the organic layer.
- Step 4: The fourth step involves the removal of the unused silicon or etching. It involves the removal of a bulk of the substrate either using wet etching or dry etching. In wet etching, the substrate is immersed in a liquid solution of a chemical etchant, which etches out or removes the exposed substrate
- Step 5: The fifth step involves the joining of two or more wafers to produce a multi-layered wafer or a 3 D structure. It can be done using fusion bonding which involves direct bonding between the layers
- Step 6: The 6th step involves assembling and integrating the MEMs device on the single silicon chip.
- Step 7: The 7th step involves the packaging of the whole assembly to ensure protection from the outer environment, proper connection to the environment, minimum electrical interference. Commonly used packages are metal can package and ceramic window package.

4.6.2. *Surface Micromachining*

Surface micromachining is a process that uses thin film layers deposited on the surface of a substrate to construct structural components for MEMS.

These manufacture gears, which are very thin, 2 to 3 microns in thickness (or height), but can be hundreds of microns wide. Each gear tooth is smaller than the diameter of a red blood cell (8 to 10 microns). These gears rotate above the surface of the substrate.

Surface micromachining uses many of the same techniques, processes, and tools as those used to build integrated circuits (ICs) or more specifically CMOS (Complementary Metal Oxide Semiconductor) components. This process is used to fabricate micro-size components and structures by depositing, patterning, and etching a series of thin film layers on a silicon substrate. This creates an ideal situation for integrating microelectronics with micromechanics.

The main difference between CMOS fabrication and surface micromachining is that the circuits constructed for CMOS allow for the movement or flow of electrons while the structures constructed with surface micromachining (e.g., cantilevers, gears, mirrors, switches) move matter.

Steps of MEMs Fabrication Using Surface Micromachining

- The first step involves the deposition of the temporary layer (an oxide layer or a nitride layer) on the silicon substrate using a low-pressure chemical vapor deposition technique. This layer provides electrical isolation.
- The second step involves the deposition of the spacer layer which can be a phosphosilicate glass, used to provide a structural base.
- The third step involves subsequent etching of the layer using the dry etching technique. Dry etching technique can be reactive ion etching where the surface to be etched is subjected to accelerating ions of the gas or vapor phase etching.
- The fourth step involves the chemical deposition of phosphorus-doped polysilicon to form the structural layer.
- The fifth step involves dry etching or removal of the structural layer to reveal the underlying layers.
- 6th step involves the removal of the oxide layer and the spacer layer to form the required structure.
- 7th step involves assembling and integrating the MEMS device on the single silicon chip.
- 8th step involves the packaging of the whole assembly to ensure protection from the outer environment, proper connection to the environment, minimum electrical interference. Commonly used packages are metal can package and ceramic window package.

4.7. Nano Electro Mechanical Systems (NEMS)

NEMS, or nanoelectromechanical systems, are devices in which the physical motion of a nanometre-scale structure is controlled by an electronic circuit, or vice versa.

Nanoelectromechanical systems (NEMS) are a class of devices integrating electrical and mechanical functionality on the nanoscale.

The Nano mechanical components are fabricated using compatible "micromachining" process.

Nano electro mechanical devices promise to revolutionize measurements of extremely small displacement and extremely weak forces, particularly at the molecular level.

NEMS devices can be so small that hundreds of them can be fit in the same space as one single micro device that performs same function.

Nano electronic integrated circuits allow nano systems to sense and control the environment.

In NEMS devices the sensors gather the information from surrounding environment through measuring mechanical chemical, biological, optical phenomenon.

The electronics then process the information derived from the sensors. Through some decision-making capability direct the actuators to respond by moving, regulating, and filtering.

There are **three basic building blocks** in NEMS technology.
- Deposition Processes
- Lithography
- Etching Process

Deposition Process

Chemical methods used NEMS deposition Process are follows

- Chemical vapor deposition
- Molecular Beam Epitaxy

Lithography

In the NEMS context is typically the transfer of a pattern to a photosensitive material by selective exposure to a radiation source such as light.

Etching

It is necessary to etch the thin films previously deposited or the substrate itself.

There are 2 class of etching process.

- Wet etching.
- Dry etching.

Wet Etching

- This is the simplest etching technology.
- There are complications since usually mask is desired to selectively etch the material.
- It requires a container with a liquid solution that dissolve the material used.
- Some single crystal material, such as silicon, exhibits anisotropic etching in certain chemicals.

Dry Etching

In dry etching process, the mask is exposed to open air using plasma or laser system to remove the unwanted material so that it can dry and clean the surface of mask. It is more expensive process.

Advantages of NEMS Devices

- Cost effectiveness
- System integration
- High precision
- Small size.

Applications of NEMS

- Wide deployment of NEMS is their use as **nano nozzles** that direct the ink in inkjet printers.
- They are also used to create miniature robots (**nano robots**) which have been rigorously tested in harsh environments for defence and aerospace where they are used as navigational gyroscopes.
- Also used as a **nano – tweezers**

Nano optical tweezers: These devices are capable of trapping and manipulating objects on the nano meter scale, even single molecules. This capability allows them to be used for a variety of applications in nano technology and micro biology.

Optical tweezers were developed in 1970 by Arthur Ashkin as a tool for the manipulation of micron-sized particles. Ashkin's original design was then adopted for a variety of purposes, such as trapping and manipulation of biological materials and the laser cooling of atoms. More recent developments have led to nano-optical tweezers, for trapping particles on the scale of only a few nanometres, and holographic tweezers, which allow for dynamic control of multiple traps in real time. These alternatives to conventional and optical tweezers have made it possible to trap single molecules and to perform a variety of studies on them. Presented here is a review of recent developments in nano-optical tweezers and their current and future applications.

NEMS in Wireless

A 3G smart phone will require the functionality of as many as five radios – for TDMA, CDMA, 3G, BLUETOOTH, and GSM operation. A huge increase in component count is required to accomplish this demand.

Thermal Actuator

Thermal Actuator is one of the most important NEMS devices, which is able to deliver a large force with large displacement.

Figure 4.4: Thermal Actuator

Overview of NEMS Devices

- Nano systems have the enabling capability and potential similar to those of nano-processors.
- Since NEMS is a nascent and synergistic technology, any new applications will emerge, expanding the markets beyond that which is currently identified or known.
- NEMS is forecasted to have growth similar to its parent IC technology.
- For a great many applications, NEMS is sure to be the technology of the future.

4.8.   Nanoelectronics: Applications Under Development

- Researchers at NIST have demonstrated an LED build with **zinc oxide nanostructures** called fins which generates much higher light output than existing designs of similar size. The researchers also found that raising the current caused the structure to generate laser light.

- Researchers at the Royal Melbourne Institute of Technology have demonstrated **atomically-thin indium-tin oxide sheets** that may make touchscreens that are cost less to manufacture and well as being flexible and consumes less power.

- Cadmium selenide nanocrystals deposited on plastic sheets have been shown to form **flexible electronic circuits**. Researchers are aiming for a combination of flexibility, a simple fabrication process and low power requirements.

- Integrating **silicon nano photonics** components into CMOS integrated circuits. This optical technique is intended to provide higher speed data transmission between integrated circuits than is possible with electrical signals.

- Researchers   at UC   Berkeley have demonstrated a low power method to use **nanomagnets as switches**, like transistors, in electrical circuits. Their method might lead to electrical circuits with much lower power consumption than transistor-based circuits.

- Researchers at Georgia Tech, the University of Tokyo and Microsoft Research have developed a method to print prototype circuit boards using standard inkjet printers. **Silver nanoparticle ink** was used to form the conductive lines needed in circuit boards.

- Researchers at Caltech have demonstrated a **laser that uses a nanopatterned silicon surface** that helps produce the light with much tighter frequency control than previously achieved. This may allow much higher data rates for information transmission over fibre optics.

- Building **transistors from carbon nanotubes** to enable minimum transistor dimensions of a few nanometers and developing techniques to manufacture integrated circuits built with **nanotube transistors**.

- Researchers at Stanford University have demonstrated a method to make functioning **integrated circuits using carbon nanotubes**. In order to make the circuit work they developed methods to remove metallic nanotubes, leaving only semiconducting nanotubes, as well as an algorithm to deal with misaligned nanotubes. The demonstration circuit they fabricated in the university labs contains 178 functioning transistors.

- Developing a lead-free solder reliable enough for space missions and other high stress environments using **copper nanoparticles**.

- Using electrodes made from **nanowires that would enable flat panel displays to be flexible** as well as thinner than current flat panel displays.

- Using **semiconductor nanowires** to build transistors and integrated circuits.

- Transistors built in **single atom thick graphene film** to enable very high-speed transistors.

- Researchers have developed an interesting method of **forming PN junctions, a** key **component of transistors, in graphene**. They patterned the p and n regions in the substrate. When the graphene film was applied to the substrate electrons were either added or taken from the graphene, depending upon the doping of the substrate. The researchers believe that this method reduces the disruption of the graphene lattice that can occur with other methods.

- Combining gold nanoparticles with organic molecules to create a transistor known as a **NOMFET (Nanoparticle Organic Memory Field-Effect Transistor)**.

- Using carbon nanotubes to direct electrons to illuminate pixels, resulting in a lightweight, millimetre thick **"nano emissive" display panel**.

- Making integrated circuits with features that can be measured in nanometers (nm), such as the process that allows the production of integrated circuits with **22 nm wide transistor gates**.

- Using nanosized magnetic rings to make **Magneto resistive Random Access Memory (MRAM)**.

- Researchers have developed lower power, higher density method using nanoscale magnets called **magnetoelectric random access memory (MeRAM)**.

- Developing **molecular-sized transistors** which may allow us to shrink the width of transistor gates to approximately one nm which will significantly increase transistor density in integrated circuits.

- Using **self-aligning nanostructures to manufacture nanoscale integrated circuits.**

- Using nanowires to build **transistors without p-n junctions**.

- Using buckyballs to build **dense, low power memory devices.**

- Using **magnetic quantum dots** in spintronic semiconductor devices. Spintronic devices are expected to be significantly higher density and lower power consumption because they measure the spin of electronics to determine a 1 or 0, rather than measuring groups of electronics as done in current semiconductor devices.

- Using nanowires made of an alloy of iron and nickel to create dense memory devices. By applying a current magnetized sections along the length of the wire. As the magnetized sections move along the wire, the data is read by a stationary sensor. This method is called **race track memory.**

- Using **silver nanowires embedded in a polymer** to make conductive layers that can flex, without damaging the conductor.

- IMEC and Nantero are developing a **memory chip that uses carbon nanotubes**. This memory is labelled NRAM for Nanotube-Based Non-volatile Random Access Memory and is intended to be used in place of high-density Flash memory chips.

- Researcher have developed an **organic nano glue** that forms a nanometer thick film between a computer chip and a heat sink. They report that using this nano glue significantly increases the thermal conductance between the computer chip and the heat sink, which could help keep computer chips and other components cool.

- Researchers at Georgia Tech, the University of Tokyo and Microsoft Research have developed a method to print prototype circuit boards using standard inkjet printers. **Silver nanoparticle ink** was used to form the conductive lines needed in circuit boards.

UNIT - V

APPLICATIONS

5.1. Nano Coatings

Nano coatings is a process by which a thin layer of thickness about <100nm is deposited on substrate for improving some property or for imparting new functionalities such as corrosion protection, water and ice protection, friction reduction, antifouling, and antibacterial properties, self-cleaning, heat and radiation resistance, and thermal management.

Nano coatings offer significant benefits for applications in the aerospace, defence, medical, marine, and oil industries, have driven manufacturers to incorporate multi-functional coatings in their products.

Nano coatings emit much lower levels of volatile organic compounds (VOCs) than traditional polymer coatings, which has been driving increased demand. They are particularly useful in contexts where there are strict VOC limits.

The Conventional Coating is Having Some Limitations Like

- Improper adhesion between coating and substrate
- Less flexibility
- Strength loss
- Poor abrasion and resistance
- Less durability.

In order to overcome these limitations, the nano coatings are developed and implemented.

Types and Applications

Anti-corrosive coatings: When applied to a metal, the coating stops chemical compounds meeting corrosive materials, this stops processes like oxidation.

Waterproof and non-stick clothing: A hydrophilic coating can be applied to various pieces of clothing, whilst the non-stick has applications in furniture, electricals and glass.

Antibacterial coating: These coatings help to inhibit the growth of microorganisms, which is particularly suitable for areas such as public transport.

Thermal barrier coating: This type of coating is particularly prevalent in the aviation industry and is normally applied to metallic surfaces. The elevated temperatures that planes work at have opened up the possibility of the coatings' use in high powered automobiles.

Anti-abrasion coatings: The main application of this coating is to prolong the life cycle of the surface by lessening the amount of friction that occurs.

Self-healing coatings: The filled nano-capsules inside this coating help repair the surface should any scratching occur. They can be found in everyday items including phones and automotive paints.

Anti-reflection coatings: This coating does not increase transmission, rather it just reduces the reflection on the incident side

Anti-graffiti coatings: These are invisible to the naked eye and prevent costly expenditure to the government and companies in cleaning up graffiti.

Advantages

- Easy to clean
- Sustainable (25-year lifetime)
- Breathable
- Resistance

Nano coatings are ultra-thin layers or chemical structures that are built upon surfaces by a variety of methods.

- Solgel
- Layer by layer (LBL)
- Soft assembly
- Dip coating
- Spin coating
- Plasma or ion-beam assisted technique
- Electro chemical deposition
- Vapour deposition
- Pulsed laser deposition
- Magnetron sputtering

Nano coating is coating that measured in nano scale i.e., 1-100nm thickness. To compare, the typical automotive paint is approximately 125 microns or 125,000 nanometers. At the nanoscale, quantum physics comes into play. Nano coatings can be structured one molecule thick, or they can be built up from multiple molecular layers. Then these are hybrid nanocoating's, such as NanoSlic, which combine multiple layers of nano coatings to deliver a wide range of benefits.

Hybrid nano coatings: such as NanoSlic Technology, which combine multiple layers of nano coatings to deliver a wide range of benefits. In these coatings, dense network of strong chemical bonds on the substrate an inert, high-performance binder polymer layer; and a hydrophobic, oleophobic contact surface. They make cleaning easier, protect and enhance surfaces.

When nano coatings are properly formulated, they can prevent finger print from forming an automotive surface and prevent harmful bacteria from growing in medical settings to helping clothing repel moisture.

They are typically transparent, nano coatings useful in applications when 'opacity' would be a problem. Opacity means, the quality or state of a body that makes it impervious to the rays of light. For ex. Certain nano coatings make windows resistant to heat and UV rays, the window stays clear but gains additional properties.

5.2. Optoelectronic Devices

Optoelectronics is the communication between optics and electronics which includes the study, design and manufacture of a hardware device that converts electrical energy into light and light into energy through semiconductors. This device is made from solid crystalline materials which are lighter than metals and heavier than insulators. Optoelectronics device is basically an electronic device involving light. This device can be found in many optoelectronics applications like military services, telecommunications, automatic access control systems and medical equipment.

This academic field covers a wide range of devices including LEDs and elements, image pick up devices, information displays, optical communication systems, optical storages, and remote sensing systems, etc. Examples of optoelectronic devices include telecommunication laser, blue laser, optical fibre, LED traffic lights, photo diodes and solar cells. Majority of the optoelectronic devices (direct conversion between electrons and photons) are LEDs, laser diodes, photo diodes and solar cells.

Types of Optoelectronics Devices

Optoelectronics are classified into different types such as

- Photodiode
- Solar Cells
- Light Emitting Diodes
- Optical Fibre
- Laser Diodes

Photo Diode

A photo diode is a semiconductor light sensor that generates a voltage or current when light falls on the junction. It consists of an active P-N junction, which is operated in reverse bias. When a photon with plenty of energy strikes the semiconductor, an electron or hole pair is created. The electrons diffuse to the junction to form an electric field.

This electric field across the depletion zone is equal to a negative voltage across the unbiased diode. This method is also known as the inner photoelectric effect. This device can be used in three modes: photovoltaic as a solar cell, forward biased as an LED and reverse biased as a photo detector. Photodiodes are used in many types of circuits and different applications such as cameras, medical instruments, safety equipment, industries, communication devices and industrial equipment.

Solar Cells

A solar cell or photo-voltaic cell is an electronic device that directly converts sun's energy into electricity. When sunlight falls on a solar cell, it produces both a current and a voltage to produce electric power. Sunlight, which is composed of photons, radiates from the sun. When photons hit the silicon atoms of the solar cell, they transfer their energy to lose electrons; and then, these high-energy electron flow to an external circuit.

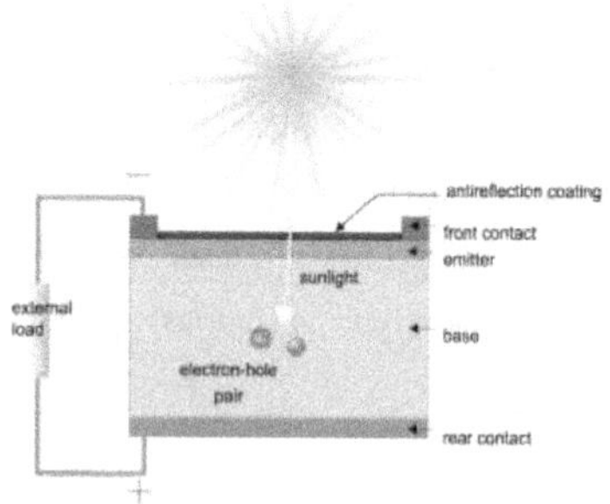

Figure 5.1: Solar Cells

The solar cell is composed of two layers which are struck together. The first layer is loaded with electrons, so these electrons are ready to jump from the first layer to the second layer. The second layer has some electrons taken away, and therefore; it is ready to take more electrons. The advantages of solar cells are that, there is no fuel supply and cost problem. These are very dependable and require little maintenance.

The solar cells are applicable in rural electrification, telecommunication systems, ocean navigation aids, electric power generation system in space and remote monitoring and control systems.

Light-emitting Diodes

Light-emitting diode is a P-N semiconductor diode in which the recombination of electrons and holes yields a photon. When the diode is electrically biased in the forward direction, it emits incoherent narrow spectrum light. When a voltage is applied to the leads of the LED, the electrons recombine with the holes within the device and release energy in the form of photons. This effect is called as electroluminescence. It is the conversion of electrical energy into light. The color of the light is decided by the energy band gap of the material.

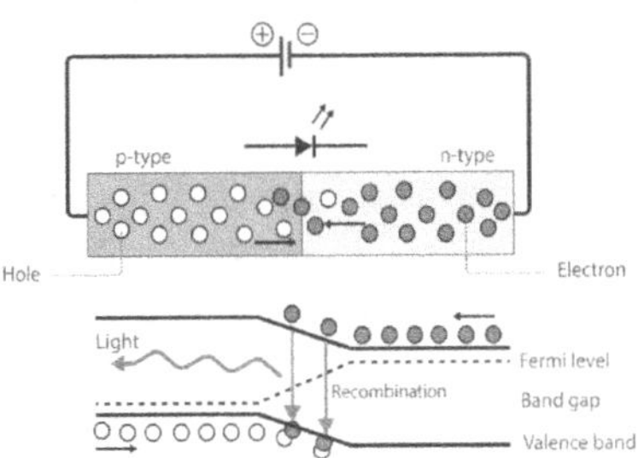

Figure 5.2: Light Emitting Diode

The usage of LED is advantageous as it consumes less power and produces less heat. LEDs last longer than incandescent lamps. LEDs could become the next generation of lighting and used anywhere like in indication lights, computer components, medical devices, watches, instrument panels, switches, fibre-optic communication, consumer electronics, household appliances, etc.

Optical Fibre

An optical fibre or optic fibre is a plastic and transparent fibre made of plastic or glass. It is somewhat thicker than a human hair. It can function as a light pipe or waveguide to transmit light between the two ends of the fibre. Optical fibres usually include three concentric layers:

a core, a cladding, and a jacket. The core, a light transmitting region of the fibre, is the central section of the fibre, which is made of silica. Cladding, the protective layer around the core, is made of silica. This creates an optical waveguide that limits the light in the core by total reflection at the interface of the core-cladding. Jacket, the non-optical layer around the cladding, typically consists of one or more layers of a polymer that protect the silica from the physical or environmental damage.

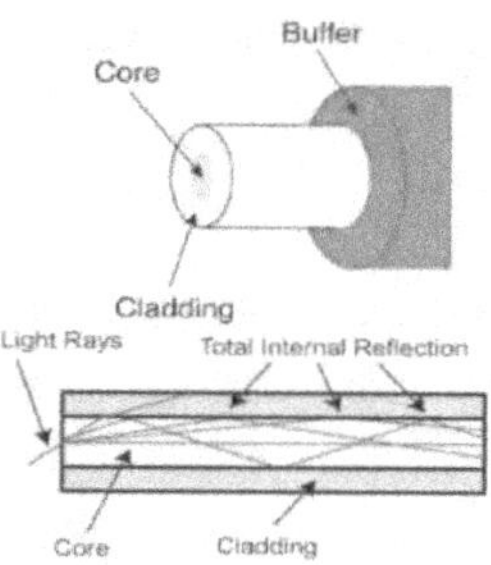

Figure 5.3: Optical Fibre

Along with the fibre-optic cable, jackets are available in different colours. These colours allow the recognition of the fibre-optic cable and the type of cable one is dealing with. For example, an orange-colour cable clearly indicates a single-mode fibre, while a yellow one indicates a multimode fibre. In the single-mode fibre, one mode propagates and the light rays travel straight through the cable. In a multimode cable, the light rays travel through the cable following different modes.

These cables are used in telecommunications, sensors, fibre lasers, bio-medicals and in many other industries. The advantages of using optical-fibre cable include their higher bandwidth, less signal degradation, weightlessness and thinness than a copper wire, cost-effectiveness, flexibility, and hence they are used in medical and mechanical imaging systems.

Laser Diodes

Laser (light amplification by stimulated emission of radiation) is a source of highly monochromatic, coherent and directional light. It operates under stimulated emission condition. The function of a laser diode is to convert electrical energy into light energy like infrared diodes or LEDs. The beam of a typical laser has 4×0.6mm extending at 15 meters. The most common lasers used are injection lasers or semiconductor lasers. The semiconductor laser changes from other lasers like solid, liquid and gas lasers.

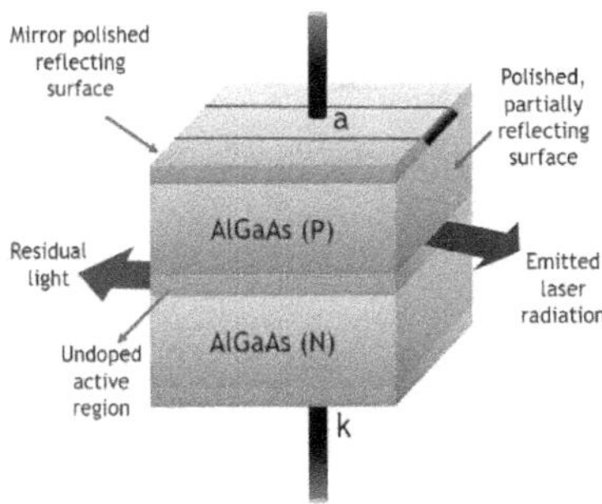

Figure 5.4: Laser Diodes

When a voltage is applied across the P-N junction, the population inversion of the electrons is produced, and then the laser beam is available from the semiconductor region. The ends of the P-N junction of the laser diode have polished surface, and hence, the emitted photons reflect back to create more electron pairs. Thus, the photons generated will be in phase with the previous photons.

Applications of Optoelectronics Devices

LEDs could become the next generation of lighting and used anywhere like in indication lights, computer components, medical devices, watches, instrument panels, switches, fibre-optic communication, consumer electronics, household appliances, traffic signals, automobile brake lights, 7 segment displays and inactive displays, and also used in different electronic and electrical engineering projects such as

- Propeller Display of Message by Virtual LEDs
- LED Based Automatic Emergency Light
- Mains Operated LED Light
- Display of Dialled Telephone Numbers on Seven Segment Display
- Solar Powered Led Street Light with Auto Intensity Control

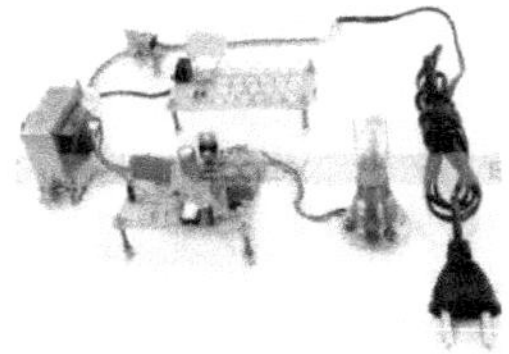

Figure 5.5: Mains Operated LED

The solar cells are applicable in rural electrification, telecommunication systems, ocean navigation aids, and electric power generation in space and remote monitoring and control systems and also used in different solar energy-based projects such as

- Solar Energy Measurement System
- Arduino based Solar Street Light
- Solar Powered Auto Irrigation System
- Solar Power Charge Controller
- Sun Tracking Solar Panel

Figure 5.6: Solar based Project

Photodiodes are used in many types of circuits and different applications such as cameras, medical instruments, safety equipment, industries, communication devices and industrial equipment. Optical fibres are used in telecommunications, sensors, fibre lasers, bio-medicals and in many other industries.

The laser diodes are used in fibre optic communication, optical memories, military applications, CD players, surgical procedures, Local Area Networks, long distance communications, optical memories, fibre optic communications and in electrical projects such as RF Controlled Robotic Vehicle with Laser Beam Arrangement and so on.

Thus, this is all about the optoelectronic devices which include laser diodes, photo diodes, solar cells, LEDs, optical fibres. These optoelectronic devices are used in different electronic project kits as well as in telecommunications, military services and in medical applications.

5.3. Environmental Nanotechnology

Nanotechnology is being used in several applications to improve the environment. This includes cleaning up existing pollution, improving manufacturing methods to reduce the generation of new pollution, making alternative energy sources more cost effective.

The Application of Nanotechnology to Environmental Issues

In trying to help our ailing environment, nanotechnology researchers and developers are pursuing the following avenues:

- **Generating less pollution during the manufacture of materials:** One example of this is how researchers have demonstrated that the use of silver nanoclusters as catalysts can significantly reduce the polluting by products generated in the process used to manufacture propylene oxide. Propylene oxide is used to produce common materials such as plastics, paint, detergents, and brake fluid.

- **Producing solar cells that generate electricity at a competitive cost.** Researcher have demonstrated that an array of silicon nanowires embedded in a polymer results in low cost but high efficiency solar cells. This, or other efforts using nanotechnology to improve solar cells, may result in solar cells that generate electricity as cost effectively as coal or oil.

- **Increasing the electricity generated by windmills.** Epoxy containing carbon nanotubes is being used to make windmill blades. The resulting blades are stronger and lower weight and therefore the amount of electricity generated by each windmill is greater.

- **Cleaning up organic chemicals polluting groundwater.** Researchers have shown that iron nanoparticles can be effective in cleaning up organic solvents that are polluting groundwater. The iron nanoparticles disperse throughout the body of water and decompose the organic solvent in place. This method can be more effective and cost significantly less than treatment methods that require the water to be pumped out of the ground.

- **Cleaning up oil spills.** Using photocatalytic copper tungsten oxide nanoparticles to break down oil into biodegradable compounds. The nanoparticles are in a grid that provides high surface area for the reaction, is activated by sunlight, and can work in water, making them useful for cleaning up oil spills.

- **Clearing volatile organic compounds (VOCs) from air.** Researchers have demonstrated a catalyst that breaks down VOCs at room temperature. The catalyst is composed of porous manganese oxide in which gold nanoparticles has been embedded.

- **Reducing the cost of fuel cells.** Changing the spacing of platinum atoms used in a fuel cell increases the catalytic ability of the platinum. This allows the fuel cell to function with about 80% less platinum, significantly reducing the cost of the fuel cell.

- **Storing hydrogen for fuel cell powered cars**. Using graphene layers to increase the binding energy of hydrogen to the graphene surface in a fuel tank results in a higher amount of hydrogen storage and a lighter weight fuel tank. This could help in the development of practical hydrogen-fuelled cars.

5.4. Nano Technology in Medicine

The use of nanotechnology in medicine offers some exciting possibilities. Some techniques are only imagined, while others are at various stages of testing, or being used today.

Nanotechnology in medicine involves applications of nanoparticles currently under development, as well as longer range research that involves the use of manufactured nano-robots to make repairs at the cellular level (sometimes referred to as *nanomedicine*).

Nanotechnology in Medicine Application

- **Drug Delivery**
  - One application of nanotechnology in medicine currently being developed involves employing nanoparticles to deliver drugs, heat, light, or other substances to specific types of cells (such as cancer cells). Particles are engineered so that they are attracted to diseased cells, which allows direct treatment of those cells. This technique reduces damage to healthy cells in the body and allows for earlier detection of disease.
  - For example, researchers at North Carolina State University are developing a method to deliver cardiac stem cells to damaged heart tissue. They attach nanovesicles that are attracted to an injury to the stem cells to increase the amount of stem cells delivered to an injured tissue.

- **Diagnostic Techniques**
  - Researchers John Hopkins University are using nano imprint lithography to Manufacture a sensor that can detect covid-19 and other viruses that can be used with hand held testing device for quick results.
  - Researchers at Worcester Polytechnic Institute are using antibodies attached to carbon nanotubes in chips to detect cancer cells in the blood stream. The researchers believe this method could be used in simple lab tests that could provide early detection of cancer cells in the bloodstream.
  - A test for early detection of kidney damage is being developed. The method uses gold nanorods functionalized to attach to the type of protein generated by damaged

kidneys. When protein accumulates on the nanorod the colour of the nanorod shifts. The test is designed to be done quickly and inexpensively for early detection of a problem.

- **Antibacterial Treatments**
  - o Researchers at the University of Houston are developing a technique to kill bacteria using gold nanoparticles and infrared light. This method may lead to improved cleaning of instruments in hospital settings.
  - o Researchers at the University of Colorado Boulder are investigating the use of quantum dots to treat antibiotic resistant infections.

- **Wound Treatment**
  - o Researchers at the University of Wisconsin have demonstrated a bandage that applies electrical pulses to a wound using electricity produced by nanogenerators worn by the patient.
  - o For trauma patients with internal bleeding another way to reduce the blood loss is needed. Researchers at Chase Western Reserve University are developing polymer nanoparticles that act as synthetic platelets. Lab tests have shown that injection of these synthetic platelets significantly reduces blood loss.

- **Cell Repair**

Nanorobots could be programmed to repair specific diseased cells, functioning in a similar way to antibodies in our natural healing processes.

- **Nanorobots in Medicine**

Future applications of nanomedicine will be based on the ability to build nanorobots. In the future these nanorobots could be programmed to repair specific diseased cells, functioning in a similar way to antibodies in our natural healing processes.

- **Nanorobots in Medicine: Future Applications**

The elimination of bacterial infections in a patient within minutes, instead of using treatment with antibiotics over a period of weeks.

The ability to perform surgery at the cellular level, removing individual diseased cells and even repairing defective portions of individual cells.

Significant lengthening of the human lifespan by repairing cellular level conditions that cause the body to age.

Table 5.1 Medical Companies

Nanomedicine Company	Product
Cristal Therapeutics	Polymeric micelle nanoparticles to deliver drugs to tumours
CytImmune	Gold nanoparticles for targeted delivery of drugs to tumours
DNA Medicine Institute	Diagnostic testing system
Mag Array	Diagnostic testing system
Blue Willow	Nanoemulsions for nasal delivery to fight viruses (such as the flu and colds) or through the skin to fight bacteria
Nanobiotix	Nanoparticles that target tumor cells, when irradiated by xrays the nanoparticles generate electrons which cause localized destruction of the tumor cells.
Nanoprobes	Gold nanoparticles for radiation therapy enhancement
Nanospectra	AuroShell particles (nanoshells) for thermal destruction of cancer tissue
Nano Viricides	Drugs called nano viricides™ designed to attack virus particles
Parvus Therapeutics	Nanoparticles designed to fight autoimmune diseases
Selecta Biosciences	Nanoparticle based synthetic vaccines
Sirnaomics	Nanoparticle enhanced techniques for delivery of siRNA
SignaBlok	Targeted delivery of drugs and imaging agents
Starpharma	Dendrimer nanoparticles for use in drug delivery
T2 Biosystems	Diagnostic testing using magnetic nanoparticles
Taiwan Liposome	Drug delivery using lipsomes
Z-Medica	Medical gauze containing aluminosilicate nanoparticles which help blood clot faster in open wounds.

5.5. Nanotechnology in Energy Generation

Some interesting ways that are being explored using nanotechnology to produce more efficient and cost-effective energy:

- **Generating steam from sunlight.** Researchers have demonstrated that sunlight, concentrated on nanoparticles, can produce steam with high energy efficiency. The "solar steam device" is intended to be used in areas of developing countries without electricity for applications such as purifying water or disinfecting dental instruments.

- o Another research group is developing nanoparticles intended to use sunlight to generate steam for use in running powerplants.
- **Generating hydrogen from sea water.** Researchers at the University of Central Florida have demonstrated the use of a nanostructured thin film of nickel selenide as a catalyst for the electrolysis of hydrogen from sea water.
- **Producing high efficiency light bulbs.** A nano-engineered polymer matrix is used in one style of high efficiency light bulbs. The new bulbs have the advantage of being shatterproof and twice the efficiency of compact fluorescence light bulbs. Other researchers developing high efficiency LED's using arrays of nano-sized structures called plasmonic cavities.
  - o Another idea under development is to update incandescent light bulbs by surrounding the conventional filament with crystalline material that converts some of the waste infrared radiation into visible light.
- **Increasing the electricity generated by windmills.** An epoxy containing carbon nanotubes is being used to make windmill blades. Stronger and lower weight blades are made possible using nanotube-filled epoxy. The resulting longer blades increase the amount of electricity generated by each windmill.
- **Generating electricity from waste heat.** Researchers have used sheets of nanotubes to build thermo cells that generate electricity when the sides of the cell are at different temperatures. These nanotube sheets could be wrapped around hot pipes, such as the exhaust pipe of your car, to generate electricity from heat that is usually wasted.
- **Storing hydrogen for fuel cell powered cars.** Researchers have prepared graphene layers to increase the binding energy of hydrogen to the graphene surface in a fuel tank, resulting in a higher amount of hydrogen storage and therefore a lighter weight fuel tank. Other researchers have demonstrated that sodium borohydride nanoparticles can effectively store hydrogen.
- **Reducing the energy used for heating and cooling buildings.** Researchers have demonstrated a device with heat absorbing sheet composed of zinc-copper nanoparticles on a thin copper layer and a heat reflecting sheet using a thin silver film. The idea is to use these heat absorbing and reflecting materials to supplement existing HVAC systems and reduce the energy required to heat and cool buildings.

- **Clothing that generates electricity**. Researchers have developed piezoelectric nanofibers that are flexible enough to be woven into clothing. The fibres can turn normal motion into electricity to power your cell phone and other mobile electronic devices.

- **Reducing friction to reduce the energy consumption**. Researchers have developed lubricants using inorganic buckyballs that significantly reduced friction.

- **Reducing power loss in electric transmission wires**. Researchers at Rice University are developing wires containing carbon nanotubes that would have significantly lower resistance than the wires currently used in the electric transmission grid. Richard Smalley envisioned the use of nanotechnology to radically change the electricity distribution grid. Smalley's concept these upgraded transmission wires, which could transmit electricity thousands of miles with insignificant power losses, with local electricity storage capacity in the form of batteries in each building that could store power for 24 hours use.

- **Reducing the cost of solar cells.** Companies have developed nanotech solar cells that can be manufactured at significantly lower cost than conventional solar cells. Check out our Nanotechnology in Solar Cells page for the details.

- **Improving the performance of batteries.** Companies are currently developing batteries using nanomaterials. One such battery will be as good as new after sitting on the shelf for decades. Another battery can be recharged significantly faster than conventional batteries.

- **Improving the efficiency and reducing the cost of fuel cells.** Nanotechnology is being used to reduce the cost of catalysts used in fuel cells. These catalysts produce hydrogen ions from fuel such as methanol. Nanotechnology is also being used to improve the efficiency of membranes used in fuel cells to separate hydrogen ions from other gases, such as oxygen.

- **Making the production of fuels from raw materials more efficient.** Nanotechnology can address the shortage of fossil fuels, such as diesel and gasoline, by making the production of fuels from low grade raw materials economical. Nanotechnology can also be used to increase the mileage of engines and make the production of fuels from normal raw materials more efficient.

www.ingramcontent.com/pod-product-compliance
Lightning Source LLC
Chambersburg PA
CBHW050543160726
48003CB00002B/736

समर्पण

यह पुस्तक मेरे माता-पिता, परिवार एवं हर उस देशवासी को समर्पित है जो भारतीय सभ्यता, संस्कृति तथा भाषाओं का हृदय से निष्ठापूर्वक सम्मान करता है। जो पूरी संवेदनशीलता के साथ, मानवता की रक्षा करते हुए प्रेम भाव से जीना जानता है।

हृदयोक्ति

मुझे न दिशाएँ समझ आतीं हैं और न रास्ते, पर फिर भी मैं नई राहों से गुजरने की हिम्मत जुटा पाती हूँ। पहाड़ की ऊँची चोटी मुझे उतनी प्रभावित नहीं करती जितनी कि वहाँ पहुँचने से पहले की यात्रा सुहाती है। वहाँ मैं कई मर्तबा रुक-रुककर न केवल अपनी श्वाँस को सामान्य करती हूँ बल्कि उस पल भर के ठहराव को भी खुलकर महसूस करती हूँ। प्रायः अपने कैमरे में कैद भी कर लिया करती हूँ। यहाँ रोज चढ़ने-उतरने वाले लोग जब डग भरते हुए आगे निकल जाते हैं तो मेरा मन उनके प्रति आदर से भर उठता है और उनके होठों के इर्दगिर्द उभरती दो लकीरें मुझमें अपार ऊर्जा का संचार कर देती हैं। रोज चढ़ने-उतरने वाले इन लोगों के मन में कभी भी इस काम को लेकर उबाऊपन नहीं दिखता। ये खुश हैं अपने-आपसे। आज के दौर में मुस्कुराते चेहरे दिखते ही कितने हैं! न जाने हँसी और प्रेम को भूल लोग व्यर्थ के तनाव और ईर्ष्या को क्यों गले लगा बैठे हैं। हर बीते पल के साथ जीवन हाथ छोड़ता जा रहा है फिर भी कुछ लोग साथ की महत्ता नहीं समझ सके!

समंदर के साथ-साथ मीलों चलना चाहती हूँ ये जाने बिना कि न जाने उस आखिरी छोर पर क्या होगा, कुछ होगा भी या नहीं! पर मैं उस तक पहुँचना चाहती हूँ। मछुआरे जाल फेंकते हैं, उनका समूह एक साथ गाते हुए एक-दूसरे का उत्साह बढ़ाता है। कभी नाव को किनारे लगाते समय सब पंक्तिबद्ध खड़े होकर रस्सी खींचते हैं। मैं ठिठककर उनके पास खड़ी हो सहायता करने की सोचती हूँ, ये जानते हुए भी कि इस रस्सी को थाम लेने भर से मैं इसे खींच नहीं पाऊँगी और यह प्रयास भी बेहद बचकाना है पर ऐसा करना अच्छा लगता है मुझे। क्योंकि उस समय उनके चेहरों पर जीवन राग की सबसे सुन्दर तस्वीर दिखाई देती है।

रेगिस्तान की भी अपनी अलग ही अदा है! यहाँ रेत के सुनहरे टीले जैसे किसी वृक्ष की तरह बाँहें फैलाए 'आओ' कहकर अपनी ओर बुलाते हैं और दूर तक पसरी मृगतृष्णा किसी नटखट बच्चे-सा खेल खेलने लगती है। एक नन्हा पौधा भी अपार प्राण वायु का अनुभव करा जाता है और उसी के नीचे खेलता वो गोल मटोल लाल-काला सा कीड़ा किसी पुष्प की तरह मन को खिला देता है। रेगिस्तान के जहाज धीरे-धीरे तसल्ली से लक्ष्य की ओर बढ़ते हैं। इन्हें किसी से होड़ नहीं

होती! बबलू, राजू, जेम्स बांड जैसे नाम धर ये झट से पर्यटकों से उम्र भर का रिश्ता जोड़ लेते हैं जिसे हम-आप जैसे लोग अपनी यात्रा की कहानियों में बार-बार दोहराते हैं।

कई बार सोचती हूँ कि बिना उठापटक, बिना राजनीति और लड़ाई-झगड़ों से दूर इन जंगलों, पहाड़ों, नदियों और रेगिस्तान की दुनिया में कितना प्रेम भरा है, कितनी सच्चाई है, कितनी एकता है। अपेक्षाएँ कम हैं, सो दुःख भी कम ही होंगे तभी तो इनकी औसत आयु हमसे ज्यादा है। तो फिर ये हमारे शहरों को ही क्या हो गया है कि इनकी भूख ही नहीं मिटती? इंसानों को क्या हो गया है कि प्रतिदिन अमानवीयता और हिंसा की वीभत्स तस्वीरों से टीवी और अखबार पटे पड़े होते हैं! इतनी सुंदर दुनिया का रंग, फीका क्यों पड़ता जा रहा है? इंसान अपनी ही धरती का सीना क्यों चीर देते हैं? आखिर कौन लील रहा है हमें? हम स्वयं ही न?

बस यह एक ही तो जीवन मिला है हमें, उसे सलीके से जीना इतना कठिन क्यों लगता है? ईर्ष्या, द्वेष, हिंसा और आपराधिक मानसिकता से लिप्त लोग, प्रेम से जीने का महत्व कब समझेंगे? कभी समझेंगे भी या नहीं! जाने कितने प्रश्न मस्तिष्क में उभरते हैं, कुछ कह पाती हूँ कुछ भीतर ही दम तोड़ देते हैं।

लेखक की धरोहर उसके शब्द ही होते हैं। मेरे पास भी यही है। एक सकारात्मक समाज की उम्मीद, सुखी इंसान की उम्मीद, हँसते चेहरों की उम्मीद, जीवन से कभी न हारने वालों की उम्मीद, प्रेम के हर हृदय में बचे रहने की उम्मीद, प्रकृति के सदा खिले रहने की उम्मीद के साथ ही लिखती आई हूँ। भविष्य में भी इसी भाव के साथ लिखती रहूँगी। आपका स्नेह और सहयोग सदैव ही मेरा मार्ग प्रशस्त करता रहा है। आशा है, यह पुस्तक आपको निराश नहीं करेगी। आप सभी तक मेरा स्नेह पहुँचे।

इस पुस्तक के प्रकाशन के लिए प्रिय नीलू सिन्हा जी को मेरा हार्दिक धन्यवाद एवं प्रखर गूँज प्रकाशन के उज्ज्वल भविष्य के लिए अनंत शुभकामनाएं!

–प्रीति अज्ञात

अनुक्रमणिका

बची रहेगी पृथ्वी?

प्रकृति के इंद्रधनुष से सजी इस सृष्टि को जब-जब निहारते हैं, इसे दाता के रूप में ही पाते आये हैं। सदियों से यह धरती हम सबका भार सह रही है। नदी, पहाड़, झरने, वनस्पति इन सभी ने मिलकर जीवन संभव किया। सूरज ने ऊर्जा भरी सुबह देकर हमें आगे बढ़ने की प्रेरणा दी तो चाँद ने थके हुए मनुष्य को अपनी नर्म बाँहों की सुकून भरी थपकी देकर उसकी हर थकान को दूर किया। शीतल प्राणवायु, साँसों को अबाध गति प्रदान करती चली आ रही है। सदियों से अडिग खड़े पहाड़ों को देख इच्छाशक्ति को और भी दृढ़ता प्राप्त होती है एवं इनसे गुजरती हिमनदी हृदय में भीतर गजब का आत्मविश्वास भर देती है। बहते पानी की मधुर ध्वनि, किसी नवजात शिशु की किलकारी सा संगीत बन ऐसा भावविभोर कर देती है कि मनमस्तिष्क स्वस्फूर्त हो उठता है। उत्साह की हिलोरें अंगड़ाइयाँ लेने लगती हैं। इन रास्तों से गुजरते हुए हँसते-झूमते फूलों और चहकते पक्षियों का दर्शन, जीवन के प्रफुल्लित भाव और सुन्दर होने की अनुभूति का जीवंत प्रमाण बन प्रस्तुत होता है।

कभी जंगलनुमा स्थानों में भटककर देखें तो यहाँ सारे पेड़ किसी बातूनी दोस्त की तरह गले मिलते हैं। उन पर फुदकते पक्षी कुछ इस तरह से मस्ती और उल्लास से भरे लगते हैं जैसे मायके आकर लड़कियाँ चहकती, इतराती घूमती हैं। कभी-कभी तो यूँ भी महसूस होता है कि किस्से-कहानियों से निकलय वो अलादीन का जिन्न या कोई सुन्दर-सी परी भी अभी सामने आ खड़ी होगी और हाथ थामय हमें वहाँ की सबसे अनोखी मीठी झील के पास छोड़ आएगी।

वस्तुतः प्रकृति ने हर तरफ से अपने स्नेहांचल में हमें घेरे रखा है। फल-फूल से लदी झुकी हुई डालियाँ और उन पर मंडराते हुए भँवरे किसका मन नहीं मोह लेते! नदी के किनारे घंटों यूँ ही बैठे रहना, कभी आसमान में टहलते बादलों की आकृतियों को समझना तो कभी पानी में उनकी छवि निहारना, प्रकृति प्रेमियों के हृदय को अपार शीतलता प्रदान करता है।

मनुष्य जाति पर यह वो ऋण है, जो हम कभी चुका नहीं सकते लेकिन इसको सहेजकर धन्यवाद अवश्य दे सकते हैं।

दुर्भाग्यपूर्ण है कि धन्यवाद देना तो दूर, हम तो इसका उचित प्रकार से संरक्षण भी न कर सके। आज धरती का अस्तित्व खतरे में है। वायुमंडल में जो भीषण प्रतिकूल परिवर्तन हुए हैं और प्राकृतिक आपदाओं के रूप में जिसका परिणाम विश्वभर को झेलना पड़ रहा है, वह मानव के कुकृत्यों की ही देन है।

हमने तालाबों का पानी चुरा लिया, उन्हें पाटकर भवन बना लिए। रसायनों के बहाव से नदियों को विषैला कर दिया, बारिश से उसके हिस्से की माटी छीन उस ज़मीन पर सीमेंट-कंक्रीट से भर अपना पट्टा लगा लिया, पानी के तमाम जलस्रोतों को नष्ट कर दिया। जंगलों को ऊँची इमारतों की तस्वीर दे दी, कीटनाशकों के अधिकाधिक प्रयोग, मिलों, कारखानों, वाहनों से बाहर निकलने वाले धुएं तथा विषैली गैसों से वायु को प्रदूषित किया। प्रकृति का जितना और जिस हद तक दोहन कर सकते थे, किया। अब हमारे पास अपना कहने को अधिक कुछ शेष नहीं है।

ये मासूम पक्षी, जो हर वृक्ष को अपना घर समझ हमारी सुबह-शाम रोशन करते हैं। जो सुबह की पहली धूप के साथ हमारे आँगन में चहचहाहट बन उतरते हैं, जो किसी बच्चे की तरह खिड़कियों से लटक सैकड़ों करतब किया करते हैं, जो अलगनी को अपने बाग का झूला समझ दिन-रात फुदकते हैं। हमने इन्हें क्या दिया? आधुनिकीकरण और विकास के नाम पर हम इनके भी घर छीन रहे हैं। जंगलों को काटकर हमने न जाने कितने जीव-जंतुओं को बेघर कर दिया। समाचार में देखकर चिंता व्यक्त करते हैं कि 'जानवर शहर में घुस आया!' नहीं! वो उसका ही घर था। उनकी ज़मीन पर अतिक्रमण के दोषी हम हैं। जब जंगल लगातार कटते जायेंगे तो कहाँ जायेंगे ये जीव-जन्तु? इनके बिना ये दुनिया कितनी सन्नाटे भरी होगी, इसकी कल्पनामात्र सिहरा देने के लिए काफी है। इनकी प्रजातियाँ यूँ ही विलुप्त नहीं हो रहीं। अपितु मानवीय स्वार्थ की बलि चढ़ रही हैं। पारिस्थितिक तंत्र के इस असंतुलन ने ही पर्यावरण सम्बन्धी समस्त समस्याओं को जन्म दिया है।

इस हरियाली का जीवन कितना निःस्वार्थ है। हरीतिमा से आच्छादित, फूल-फल से लदे वृक्ष बिन कहे ही सकारात्मक सन्देश दे जाते हैं। ये मुसाफिरों को छाया देते हैं, अपनेपन का एहसास कराते हैं, बिन किराए, कुछ पल निश्चिंतता से बैठने की सुविधा देते हैं। इन्हें अपने होने का मतलब पता है, ये शिकायत नहीं करते, ये जानते हुए भी कि कोई राहगीर पलटकर उनकी तरफ वापिस कभी नहीं आएगा। उन्हें यह भी पता है कि दिन की चटक रोशनी में ही हमें उनकी जरूरत महसूस होती है और एक दिन कुल्हाड़ी के तीक्ष्ण प्रहार से उन्हें नष्ट कर दिया जाएगा। लेकिन फिर भी उसकी शाखाएँ सदैव आलिंगन की मुद्रा में ही दिखेंगी। वृक्ष बचेंगे तो डालियों से झूलते झूले की परंपरा बनी रहेगी और बच्चों का खिलंदड़पन नित नई उड़ान भरेगा। संस्कृति जगमगा उठेगी।

पानी की कितनी कमी है और जल-स्तर भी लगातार घटता जा रहा है, इस दुखद तथ्य से हम सब भली-भाँति परिचित हैं।

लेकिन औद्योगीकरण और बढ़ती हुई जनसंख्या के लिए एक अदद घर का रोना लेकर हम इस बात को भूलते हुए लगातार जंगल काटते आ रहे हैं कि वृक्ष और मिट्टी के बिना बारिश का पानी पक्की ज़मीन कहाँ से सोखेगी? ऐसे में जल-स्तर घटना तय ही है जब उसे बढ़ाने के सारे माध्यमों का मुँह हम ही बंद कर रहे हैं। जल के प्रदूषित होने और गाँवों-शहरों में इसके निकास की समस्या और उससे पनपती बीमारियों की जो गाथा है सो अलग! यह इस भू-मंडल पर रहने वाले प्रत्येक मनुष्य का अधिकार है कि उसे पीने के लिए स्वच्छ जल मिले लेकिन विडंबना देखिये कि मनुष्य ही इसकी अनुपलब्धता के लिए उत्तरदायी भी है। दो-ढाई दशक पूर्व की ही बात करें तो हमारे देश में 'पानी के व्यवसाय' का अस्तित्त्व तक नहीं था और न ही इसके शुद्धिकरण की कोई आवश्यकता ही थी। कब धीरे-धीरे हम सबके घरों में प्यूरीफायर लग गए या बोतलबंद पानी आने लगा, पता भी नहीं चला। पानी को शुद्ध बनाने, इसकी पैकेजिंग और इसे बेचने के लिए तमाम कम्पनियाँ बनती चली गईं और इसके विज्ञापन भी सराहे जाने लगे। लेकिन मूल समस्या अब भी जस-की-तस है। विकास के नाम पर जितना शोषण इस धरती का किया गया है उसके दुष्परिणाम अब दृष्टिगोचर होने लगे हैं। आने वाली पीढ़ियों के लिए यह स्थिति जितनी भयावह होगी, इससे निबटना उतना ही दुष्कर भी रहेगा।

पर्यावरण से सामंजस्य बिठाये बिना मानव सभ्यता के विकास की कल्पना ही नहीं की जा सकती। प्रत्येक क्षेत्र विशेष में वहाँ की परिस्थितियों के अनुकूल ही वनस्पतियों का विकास तथा जीव-जंतुओं का आवास तय होता है। प्रतिकूल पर्यावरण में पादप-समूहों तथा जीवों की प्रजातियाँ विलुप्त हो जाती हैं अथवा उनकी प्राकृतिक क्रिया में अवरोध उत्पन्न होता है। इस विषम परिस्थिति का सीधा प्रभाव वहाँ की कृषि, अन्य उद्योग एवं मानवीय जीवन पर पड़ता है। जितनी लापरवाही से हम सब नष्ट करते जायेंगे उतने ही दुरूह परिणामों का मिलना भी तय है।

प्रकृति की प्रकृति में कुछ भी नहीं बदला। बस, मनुष्य ही खोटा निकला! हमें प्रकृति जैसा बनना होगा। इससे जीने की प्रेरणा लेनी होगी।

आसमान सिखाता है कि माँ के आँचल का अर्थ क्या है।

पहाड़ सच्चे दोस्त की तरह, किसी अपने के लिए डटकर खड़े रहने की बात कहते हैं।

सहनशक्ति धरती से सीखनी होगी।

प्रेम क्या होता है और बिना किसी अपेक्षा के अगाध स्नेह कैसे किया जाता है, किसी के दुख में साथ कैसे दिया जाता है, इस

कला को पशु-पक्षियों से बेहतर कोई नहीं जानता।

कट-कटकर गिरने के बाद फिर कैसे बार-बार उठना है, आगे बढ़ना है, जीवन का यह पाठ पौधे सिखाते हैं। जब तक ये जीवित हैं प्राणवायु भी देते हैं कि हमारे अस्तित्त्व को खतरा न रहे!

धूप, हवा, पानी से दान की प्रसन्नता महसूस करनी होगी।

सभ्यता, बीजों और माटी से समझनी होगी जो उगने के लिए धूप और पोषक तत्त्व साझा करते हैं।

यदि हम ऐसा कर पाते हैं तो ही मनुष्यता को पा सकेंगे। उसके पश्चात् ही प्रकृति और पर्यावरण के संरक्षित होने की आशा रख सकते हैं। सरकारें कुछ नहीं कर सकतीं जब तक कि हम मनुष्यों की सोच निज हितों से ऊपर न उठे। जिन कारणों से पर्यावरण को हानि पहुँच रही है उन क्रियाकलापों का उचित प्रबंधन होना आवश्यक है साथ ही पर्यावरण संरक्षण हेतु जनमानस को जागरूक एवं सचेत भी करना होगा।

सृष्टि ने हमें वो सब दिया जो इस धरती को स्वर्ग बना सकता था लेकिन हम उसे प्लास्टिक और विषैले पदार्थों से भर अपनी असभ्यता का परिचय देते रहे। परिस्थितियाँ इतनी विकट हो चुकी हैं कि अब वैश्विक स्तर पर यह घनघोर चिंता का विषय बन चुका है। यहाँ तक कि पृथ्वी के समाप्त होने की तिथि भी आये दिन घोषित होने लगी है।

समय तेजी से आगे बढ़ रहा है पर अभी बीता नहीं! हम सब को अपनी इस धरा से प्रेम है और जब तक ये प्रेम जीवित है तब तक पृथ्वी के बचे रहने की उम्मीद कायम है। यदि प्रेम ही खतरे में है तो फिर यूँ ही जीकर करेंगे भी क्या और किसके लिए! झरनों में जो संगीत है, नदी की जो कलकल है, पहाड़ों से लिपटी जो बर्फ है, तारों से खिलखिलाता जो आकाश है, ये बाँहें फैलाए जो वृक्ष खड़े हैं और फूलों की ये सुगंध जिसे घृणा की हजारों जंजीरें भी अब तक बाँध नहीं सकी हैं, ये पक्षी जो आसपास फुदकते, चहचहाते हैं। यही प्रेम है, यही हमारी प्यारी पृथ्वी है। यह हमारी वो विरासत है जिसे हमें आने वाली नस्लों के लिए सुरक्षित रख छोड़ना है कि वे जब इस दुनिया में आयें तो ये उन्हें भी इतनी ही सुन्दर दिखाई दे जिसके प्रत्यक्षदर्शी हम सब रहे हैं।

मनुष्य से बड़ी आपदा कोई नहीं

'प्रेम' एक ऐसा विषय है, जिससे जुड़ी कोई न कोई कहानी सबके पास होती है। आपको भी अपने विद्यालय-महाविद्यालय की कई घटनाएँ तो अवश्य ही याद होंगी। वह लम्बी सूची भी कहाँ भूल पाए होंगे, जिसे अपने मित्रों को अब भी सुनाया करते हैं कि कौन-कौन से लड़के/लड़की आप पर जान छिड़कते थे। संभवतः किसी के प्रति अपनी एकतरफा मुहब्बत को सोच आपका मन भी कभी-कभी उदास हो जाया करता होगा। इस बात की स्मृति भी आपके हृदय को झकझोरती होगी कि कैसे, किसी की बे-पनाह मुहब्बत को ठुकरा कर एक लड़के/लड़की ने किसी और का हाथ थाम लिया था।

कितने ही किस्से सुने-पढ़े होंगे, जिसमें माता-पिता की जिद के आगे प्रेमियों को झुकना पड़ा या परिवार के दवाब के चलते उनका विवाह अन्यत्र तय हो गया। हमारे समाज में माता-पिता के ऋण को चुकाने का सर्वोत्तम समय यही बताया गया है। राय दी जाती है कि भूल जाओ अपने प्रेम को, उस इंसान को दिल से सदा के लिए निकाल दो जिसे आप अपना जीवन मान बैठे हैं। जिन्होंने जीवन दिया, उनका मान रखने या इस जीवन की कीमत चुकाने का यह अचूक तरीका, हमारे इसी समाज का बनाया हुआ है। आश्चर्यजनक नहीं लगता एक के प्रेम के बदले दूसरे के प्रेम को भूल जाने का सौदा?

हर बात में 'प्रेम से रहो', 'प्रेम बाँटो', 'प्रेम से बड़ा कोई धर्म नहीं' कहने वाली यह दुनिया इस 'प्रेम' की ही बलि सबसे पहले ले लेती है। समाज और इज़्ज़त की दुहाई के नाम पर प्रेमियों को ही सबसे अधिक सताया गया है। सदियों से चला आ रहा यह क्रम, दुर्भाग्य से अब तक भी टूटा नहीं है।

प्रकृति में कोई भी प्राणी प्यार करने या निभाने के लिए किसी की अनुमति नहीं लेता, सिवाय मनुष्य के। सोचिए कोई भी पशु-पक्षी, बे-रोक-टोक किसी से भी प्यार कर सकता है, जब जी चाहे उससे मिल सकता है, उम्र भर के लिए जोड़ा बना सकता है। लेकिन दो मनुष्यों के बीच प्राकृतिक प्रेम की कहानी चर्चा का विषय बन जाया करती है। प्रेम करना माने भीषण अपराध हो गया। यदि उन्होंने किसी की परवाह किये बिना शादी की तो उसे 'भागकर की गई शादी' का टैग दे दिया जाता है। इसे यूँ समझिये कि प्रकृति प्रदत्त प्रेम, समाज से भागकर ही किया जा सकता है। दिल से समाज की नाक जो टकरा जाती है!

समाज के ठेकेदारों द्वारा नाक की कीमत, प्रेम से बहुत अधिक तय की गई है। इनकी महानता दूसरों के प्रेम में दखल देकर ही सिद्ध होती है। केवल इतना ही नहीं, इन्होंने जाति, धर्म के आधार पर भी नियम तय कर दिए और बची-खुची जिम्मेदारी परिजनों ने उठा ली। बस, प्रेम करने के नियम की सरकारी पाठ्यपुस्तक और बना देनी चाहिए, जिससे शुरू से ही सबको पता रहे कि इसके कारण आपको बीच चौराहे पर अपमानित किया जा सकता है, अपनों से जूझना पड़ सकता है और जो समाज के बनाए खाँचे में फिट न बैठे तो दुर्भाग्य से जान भी गंवानी पड़ सकती है।

हमने हर छोटी-बड़ी बात के लिए हड़ताल की, आंदोलन किए लेकिन सृष्टि के सबसे खूबसूरत भाव 'प्रेम' को बचाने के लिए हम क्या कर सके? क्या करना चाहा है? क्या प्रेम को बचाने के लिए भी कभी कोई राष्ट्रीय केम्पैन चलाया गया है? हम मनुष्य सबको सहेजने, बचाने की बड़ी-बड़ी बातें कहते हैं, हर समय संवेदनशीलता का चोगा ओढ़े घूमते हैं, प्रेम का खूब बखान करते हैं लेकिन मारने-काटने, बदला लेने की मानसिकता से अब तक बाहर नहीं निकल सके हैं। एक तरफ प्रेम की स्तुति और दूसरी तरफ घृणा का पोषण, विध्वंसता को बढ़ावा... वाह! मनुष्य के दोगलेपन की कोई सीमा ही नहीं! यदि यही हमारा प्रेम है तो फिर 'प्रेम ही पूजा है' जैसी खोखली बातें किसलिए?

क्या अब समय नहीं आ गया है 'प्रेम' को पूरी तरह से आत्मसात कर लेने का? क्या अब नहीं महसूस होता कि बेटी बचाओ, पर्यावरण बचाओ, सेव टाइगर, सेव बर्ड्स, स्त्री सम्मान के लाखों आह्वानों से भरी इस दुनिया में, जो हमने बस इक प्रेम को बचा लिया होता तो सब तरफ से यूँ बचाओ-बचाओ की पुकार न सुनाई देती! प्रेम को बचा लेने भर से सृष्टि की हर शय स्वतः ही संरक्षित हो जाती। सबके चेहरे पर मुस्कान सजी रहती क्योंकि प्रेम केवल जोड़ना जानता है। प्रेम, प्रेम ही सिखाता है। यह इंसान को ईश्वरतुल्य बना देता है। दुखद है कि इसकी शक्ति का अनुमान सबको है पर इसे जीवन में गिनती भर लोग ही उतार पाते हैं।

हम तो इतने महान हैं कि इंसानों के प्रेम की तरह ही, हमने प्रकृति को भी बाँधना चाहा जबकि प्रकृति की सुंदरता उसकी स्वच्छंदता में है। नदियों की सुंदरता उसके कल-कल बहाव में है। पहाड़ आसमान को चूमते हुए ही सुंदर दिखते हैं। जंगल की विशेषता ही यही है कि वहाँ वनस्पतियाँ एवं पशु-पक्षी निर्बाध, उन्मुक्त हो रह सकें। लेकिन बाँध बनाने के शौकीन हम कभी रुकेंगे? हमने पहाड़ काटे, जंगल काटे, पशु-पक्षियों को मारा, नदियों को सुखा दिया, प्रकृति के साथ जितनी और जिस हद तक छेड़छाड़ की जा सकती

थी, हमने की। विकास के मखमली पैकेट को दिखाकर हमने प्रकृति को रौंद कर रख दिया। आज जब बाढ़, भूकंप, ग्लेशियर का फटना जैसी घटनाओं के रूप में प्रकृति का क्रोध हम पर उफान मारता है तो हम बिलबिला उठते हैं। मतलब हम किसी के साथ सदियों तक आततायी बन मनमानी करते रहें और वो पलटकर एक तमाचा भी न जड़े!

आश्चर्य है कि जब प्रकृति की भृकुटी तनती है तो हम उसे आपदा का नाम दे देते हैं लेकिन हम स्वयं, प्रकृति के लिए आपदा बन बैठे हैं इस सच को कब स्वीकारेंगे? जो हमने प्रकृति से प्रेम किया होता, सबको उसके प्राकृतिक रूप में रहने दिया होता तो आपदा के नाम पर तमाम अपनों को यूँ असमय ही न खोया होता। हमें स्वर्ग की कल्पना भी न करनी पड़ती क्योंकि वह इस धरती पर ही दृष्टिगोचर होता।

मानव हो या प्रकृति, प्रेम ही इन्हें सँवारता है। आप प्रेम करने वालों से लाख खुन्नस रखिए पर ये तो आपका दिल भी मानता है न कि जहाँ 'प्रेम' है, वहाँ सब है। जो प्रेम नहीं तो कुछ भी शेष न रहेगा! बचा लीजिए न, इस प्रेम को!

ह्वेनसांग के आँसुओं के मायने

भारत-चीन तनाव के चलते एक पौराणिक कथा अत्यधिक प्रासंगिक लगने लगी है जब चीनी यात्री ह्वेनसांग का भारत आगमन हुआ था। यह वो समय था जब पूरा विश्व भारतीय संस्कृति से प्रभावित था।

क्या है वह किस्सा?

यह चक्रवर्ती सम्राट हर्षवर्धन के शासनकाल की घटना है। ह्वेनसांग ने भारतवर्ष के कई प्रमुख स्थानों की यात्रा की तथा वे लोगों के व्यवहार, भारतीय संस्कृति और परम्पराओं से अवगत हुए। हमारे धर्मग्रंथों और इतिहास के प्रति भी उनकी बहुत रुचि थी और उनके पास इनका अच्छा-खासा संग्रह भी था। स्वदेश लौटने से पहले उन्होंने अपने अनुभव सम्राट हर्ष के साथ साझा किये और उनके प्रति कृतज्ञता व्यक्त की। हर्षवर्धन ने भी उन्हें सम्मानित कर विविध उपहार दिए, साथ ही उनके सकुशल अपने देश पहुँचने हेतु नौका एवं बीस योद्धा सैनिकों की भी उचित व्यवस्था की।

ह्वेनसांग के प्रस्थान के समय सम्राट ने अपने योद्धाओं से कहा, 'इस नौका में कई भारतीय धर्मग्रंथ एवं ऐतिहासिक वस्तुएँ हैं जो हमारी संस्कृति का प्रतीक हैं। इनकी रक्षा करना आप सभी का कर्तव्य है'।

कई दिनों तक उनकी सुखद यात्रा चलती रही लेकिन एक दिन समुद्र में भयंकर तूफान आया और नौका डोलने लगी। सभी भयभीत हो गए। घबराकर प्रधान नाविक ने कहा, 'नौका में भार अधिक हो गया है। शीघ्र ही ये पुस्तकें एवं ऐतिहासिक वस्तुओं को समुद्र में फेंक अपने प्राणों की रक्षा कीजिए'।

यह सुनकर सैनिकों के नायक ने कहा, 'यह हमारा अग्निपरीक्षा काल है। इन सभी वस्तुओं के रक्षण द्वारा भारतीय संस्कृति की रक्षा करना हमारा प्रधान कर्तव्य है। इन्हीं ग्रंथों से तो लोगों को हमारी सभ्यता और परम्पराओं का ज्ञान होगा! इसकी रक्षा के लिए हम अपने प्राण भी समर्पित कर सकते हैं'। अपने नायक के ये वचन सुनते ही तुरंत सभी योद्धाओं ने एक साथ पानी में छलांग लगा दी।

अकस्मात हुई इस घटना से ह्वेनसांग हतप्रभ हो गए। संस्कृति की रक्षा हेतु भारतीय वीरों के त्याग और बलिदान को देख उनकी आँखों से अविरल अश्रुधार बहने लगी।

'वसुधैव कुटुंबकम' ही हमारा मूल स्वभाव है

यह मात्र भावविह्वल कर देने वाली कथा भर ही नहीं है बल्कि यह हमारी प्राचीन संस्कृति और संस्कारों की प्रतिनिधि कहानी

के रूप में भी उभरती है। यह इस तथ्य को और सुदृढ़ करती है कि जब-जब संस्कृति को अक्षुण्ण बनाए रखने के लिए अपने जीवन को भी दाँव पर लगा देने का अवसर आया है, तब-तब हमारे समर्पित वीरों ने बिना किसी हिचकिचाहट के संस्कृति को ही चुना। इसे दुर्भाग्य ही कहेंगे कि इस वीर गाथा के साक्षी उस देश के यात्री ह्वेनसांग के आँसू हैं जो देश (चीन) आज हम पर आँखें तरेरने का दुस्साहस कर रहा है।

हम भारतीय अब तक अपने उन संस्कारों को नहीं भूले हैं। हमने हर बार दुश्मन देश से आये लोगों का खुलकर स्वागत किया है और किसी भी वैमनस्यता को पीछे रख, सदैव ही प्रेमपूर्वक दोस्ती का हाथ बढ़ाया है। यह न केवल हमारे शांतिप्रिय होने की बल्कि सहज विश्वास की भी पुष्टि करता है। 'वसुधैव कुटुंबकम' हमारे मूल स्वभाव में है।

अतिथि देवो भवः

उपर्युक्त पूरी घटना यह भी सिद्ध करती है कि हमारे देश में अतिथियों को मान देने की परम्परा सदियों से चली आ रही है और आज तक कायम है। माना कि उस समय हमारा देश सोने की चिड़िया था और घर आये अतिथि को हम बेशकीमती उपहारों से लादकर ही विदा करते थे लेकिन अतिथि सत्कार का यह भाव आज भी हमारी संस्कृति का उतना ही अहम् हिस्सा है। हम उस अटूट संस्कृति के भी सम्पुष्ट वाहक रहे हैं जहाँ देशधर्म, देशहित सबसे पहले है।

'हिन्दी चीनी भाई-भाई'?

चाहे वह नेहरू जी के समय 'हिन्दी चीनी भाई-भाई' का नारा हो या शी जिनपिंग के साथ मोदी जी का झूला झूलना, यह सब हमारा भरोसा, अपनत्व भाव और पुरानी कटुता को भूलकर आगे बढ़ना ही प्रदर्शित करता है। हम हर किसी के व्यवहार को रणनीति की दृष्टि से नहीं देखते बल्कि उस पर पूरा भरोसा करते हैं। जो भी हमसे मित्रवत होकर मिला, हमने उसे गले ही लगाया है।

हम आज भी सारी कड़वाहट को पीछे छोड़ते हुए चाइनीज सामान को अपना पूरा बाजार दे देते हैं। उनके उत्पादों के लिए पलक-पाँवड़े बिछाए रखते हैं। देश भर में उनके होर्डिंग्स टांग लेते हैं।

चीन ने ह्वेनसांग के आँसुओं से क्या सीखा?

ह्वेनसांग ने अपने देश जाकर नम आँखों से इन तमाम पुस्तकों का अनुवाद किया। 'सी-यू-की' ग्रन्थ उनकी भारत यात्रा का विवरण हैं जिसमें उन्होंने भारत की आर्थिक, सामाजिक, औद्योगिक दशा एवं संस्कृति का विस्तार से वर्णन किया है। लेकिन लगता है

चीन ने ह्वेनसांग के आँसुओं से कुछ नहीं सीखा, तभी तो वह बंजर ज़मीन के टुकड़े के लिए आँखें तरेरने लगा है। उसे सीमा के विस्तार में दिलचस्पी है, भले ही इस प्रक्रिया में उनके दिल सिकुड़ते चले जाएँ!

हमारे लद्दाख में चीनियों की ऊलजलूल और फितूर भरी शर्मनाक हरकतों से इनके पूर्वजों की आत्माएँ भी शर्मिंदा होती होंगी। चीन का वर्तमान रवैया, ह्वेनसांग की भावनाओं का अपमान है, उस संवेदनशील हृदय की अवहेलना है। हमारी संस्कृति को समझने के लिए चीन को ह्वेनसांग की आँखों से हमें देखना होगा। उनसे बहे आँसुओं से सीखना होगा! विचारणीय है कि आखिर चीनियों के लिए ह्वेनसांग के आँसुओं की कीमत है ही क्या!

बुलेटप्रूफ है गांधीवाद

कैसी विडंबना है कि जिस व्यक्ति की जयंती सारी दुनिया 'अंतरराष्ट्रीय अहिंसा दिवस' के रूप में मनाती है, उनकी मृत्यु हिंसात्मक तरीके से हुई। यह गांधी जी की ही हत्या नहीं थी यह उस विचार, उस सिद्धांत की हत्या का कुत्सित प्रयास भी था जिसे गांधी ने अपने जीवन का मूल आधार बनाया था। उनके हत्यारों ने भले ही एक निर्जीव देह को देख उस समय उत्सव मना लिया हो लेकिन वे हमारे बापू को नहीं मार सके! यदि वे तीन गोलियाँ नहीं लगतीं, तो शायद गांधी जी बाद में किसी रोग या अन्य प्राकृतिक कारण से स्वर्ग सिधारते। लेकिन जिस तरह उनकी हत्या हुई, उसने अहिंसावाद को और भी अधिक गहरे से रेखांकित कर दिया। कारण स्पष्ट है कि हमारे राष्ट्रपिता महात्मा गांधी, हमारे बापू का अस्तित्व केवल मोहनदास करमचंद गांधी तक ही सीमित नहीं था। वे अपने विचारों के साथ तब भी सर्वप्रिय थे और आज भी जनमानस में बसे हैं। जैसे देह के निष्प्राण होने के पश्चात अजर–अमर, आत्मा जीवित रहती है वैसे ही गांधी भी हम सबके हृदय में विद्यमान हैं। हर सच्चे भारतीय की रग में उनके सत्य का राग प्रवाहित होता है, अहिंसा की रागिनी धड़कती है। गांधी सदैव प्रासंगिक रहेंगे क्योंकि गांधी मात्र एक व्यक्ति नहीं, सम्पूर्ण विचारधारा है।

आपने भी यह खबर सुनी होगी कि बापू के सम्मान में ब्रिटेन की सरकार ने एक सिक्का जारी करने की घोषणा की थी। निस्संदेह यह उनका सम्मान तो है ही लेकिन एक तथ्य और भी है। वो यह कि अंग्रेजी हुकूमत के अत्याचारों, प्रताड़ना और अमानवीय कुकृत्यों से इतिहास रंगा पड़ा है। इस शर्मिंदगी से बचने और स्वयं पर लगे कलंक को धोने के लिए भी यह एक आवश्यक कदम था। अमेरिका में भी एक अश्वेत नागरिक जॉर्ज फ़्लॉइड की निर्मम हत्या के बाद वहाँ की जनता ने आंदोलन प्रारंभ कर दिया था और तबसे नस्लवाद का वैश्विक विरोध तेजी पकड़ रहा है। विचारणीय है कि जहाँ आज सारी दुनिया भूल सुधार के तमाम यत्न कर रही है ऐसे में हम कहाँ हैं और किन बातों पर लड़ रहे हैं?

दुर्भाग्य है कि विश्व भर में पूजनीय और सबको सत्य, अहिंसा का पाठ पढ़ाने वाले गांधी जी का विरोध करने वाले लोग, उनके अपने ही देश में पनप रहे हैं। घृणा के बीज बो रहे हैं। एक ओर हम राष्ट्रप्रेम की माला जपते हैं और दूसरी तरफ राष्ट्रपिता को अपशब्द कहने वालों के विरुद्ध कोई कार्यवाही होती नहीं दिखती! न जाने, देश के प्रति यह किस तरह का प्रेम है जो शहीदों के अपमान को यूँ सहन कर लेता है!

आज भी ऐसे लोग मिल जाते हैं जो गांधी जी का नाम सुनते ही मुँह बिचकाकर कहते हैं कि आजादी चरखा चलाने और लाठी से नहीं मिली! हाँ, भई मानते हैं कि हमारे स्वतंत्रता सेनानियों का योगदान अतुलनीय एवं अविस्मरणीय है। देश को आजाद कराना किसी एक व्यक्ति के बस की बात हो भी नहीं सकती! महत्वपूर्ण है, लालच से दूर रह अपने उसूलों पर चलना। शांति से अपनी बात कहना, सबकी राह प्रशस्त करना। गांधी विरोधी यह भूल जाते हैं कि गांधी का व्यक्तित्व और कहन का प्रकार बहुआयामी है और जैसा वह समझ रहे यह उससे इतर और विशाल भी है।

चरखा प्रतीक है स्वावलम्बन और स्वाभिमान का, स्वदेशी उत्पाद को बढ़ावा देने का। गांधी जी अपना काम स्वयं करने में विश्वास रखते थे। स्वच्छता पसंद थे। यदि हम भारतीयों ने मात्र ये दो बातें भी आत्मसात कर ली होतीं तो आज उनके निधन के सात दशक बाद 'आत्मनिर्भर भारत' या 'स्वच्छ भारत अभियान' का राग अलापने की कोई आवश्यकता नहीं होती। इन संस्कारों की नींव बहुत पहले ही पड़ गई थी। हमारे पूर्वजों ने तो सदैव ही हमें उच्चतम जीवन मूल्यों को हस्तांतरित किया है ये हमारी मूर्खता रही कि हम ही उन्हें समय पर न सहेज सके और न ही उन्हें अपनाने की कोई सार्थक चेष्टा ही की।

साबरमती आश्रम में प्रायः गांधीवादी उम्रदराज लोगों से मिलना होता है। इनमें से कई ऐसे हैं जिन्होंने बापू पर बहुत शोधकार्य किया है। वे आँखों में चमक लिए उनकी बात करते हैं तो मन को बहुत अच्छा लगता है। कुछ चेहरों पर निराशा की झलक उदास भी कर जाती है, जब वे बुझे हृदय के साथ ये कहते हैं कि 'आज के दौर में शांति और अहिंसा की बात करना मूर्खता एवं अपना उपहास करवाने से अधिक कुछ नहीं!' मैं उनकी बातों को सुन बेहद शर्मिंदा महसूस करती हूँ कि जिस दौर की ये बात कर रहे हैं, हमने उसे क्या बना दिया!

सोचती हूँ कि जब हम अपराध के विरोध में खुलकर न बोले तो अपराधी ही तो हुए न! अन्याय को देख हमने अपनी जुबान सिल ली, तो फिर हम न्यायप्रिय कैसे हुए? झूठ को जीतते देख ताली बजाने वाले, किस मुँह से 'सत्यमेव जयते' कह सकेंगे? मैं मुस्कुराते हुए उन लोगों से बस इतना ही कहती हूँ कि 'अरे! आप चिंता न कीजिए, सब अच्छा होगा। बुरे दौर का भी अंत तय ही होता है। गांधी जी की विचारधारा हताश भले ही दिख रही है पर जीवित है अभी और सदैव रहेगी'। न जाने मैं उन्हें कह रही होती हूँ या स्वयं को ही आश्वस्त करने का प्रयास करती हूँ लेकिन हर बार बापू की मूर्ति को प्रणाम करते हुए जब उनका शांत चेहरा और मृदु मुस्कान

दिखती है तो उम्मीद फिर जवां होने लगती है। विचारों में दृढ़ता आती है और हृदय पुकार उठता है 'सत्यमेव जयते'।

विश्वास गहराने लगता है जिस व्यक्ति की निर्मम हत्या भी उसके द्वारा प्रज्ज्वलित अहिंसा की मशाल न बुझा सकी, वह हमारे बीच से भला कैसे जा सकता है कभी! सदियाँ बीतती जाएंगी लेकिन सभ्यता की नजीर बन गांधी एवं उनका बुलेटप्रूफ गांधीवाद सदा स्थापित रहेगा। उनके हत्यारों को इससे बड़ा तमाचा और क्या होगा!

स्वतंत्रता का मान

सैंतालीस से लेकर अब तक यदि हम पीछे पलटकर देखें तो भारत ने हर क्षेत्र में उन्नति की है चाहे वह शिक्षा हो, चिकित्सा हो, उद्योग, विज्ञान, सुरक्षा या कोई भी क्षेत्र, हमारे विकास की कहानी हर सफहे पर सुनहरे अक्षरों में दर्ज है। हाँ एक बात अवश्य है कि इसके लिए कोई भी दल जब श्रेय लेने या दूसरे के काम में कमी निकालने पर आमादा हो जाता है तो वह दृश्य अत्यधिक वीभत्स एवं अमानवीय लगता है। जनता पूरे विश्वास के साथ अपने नेताओं को चुनती ही इस उम्मीद के साथ है कि वे देश की प्रगति में योगदान देंगे साथ ही उनके लिए भी कुछ सकारात्मक करेंगे। लेकिन न जाने क्यों राजनीति उसे सदा ही अहसान की तरह परोसती आई है।

दुनिया के सामने हमारी प्रगति तो बहुत हुई है और इस हद तक हुई है कि हम अपनी भारतीयता पर गर्व कर नित इस धरती का माथा चूम सकते हैं। दुःख यह है कि एक तरफ जहाँ हम भौतिक विकास यात्रा की ओर बढ़ते जा रहे हैं तो वहीं दूसरी ओर हमारी मानसिकता अत्यंत ही दूषित हो चुकी है। जिस स्वतंत्रता ने हमें पंख दिए और बाँहें फैलाए आसमान तक पहुँचने की उड़ान भरने का हौसला दिया, हमने उससे दूसरों को घायल करना पहले सीख लिया।

हम अपनी लकीर बड़ी करने के स्थान पर दूसरों की छोटी करने और उसे मिटाने पर अधिक जोर देने लगे हैं।

हमने इतिहास को पढ़ तो लिया पर उसे ठीक से समझा नहीं और न ही उसकी गलतियों से कोई सीख ही ली। यही दुर्दशा कट्टरपंथियों की भी है जो समय के साथ आगे बढ़ना ही नहीं चाहते। इन्हें हर बात पर आग उगलना आता है। भले ही उसकी लपेट में मानवता त्राहिमाम कर उठे!

हमें विचार करना होगा कि-

अभिव्यक्ति के नाम पर हम कहीं अपने ही अधिकारों का दुरुपयोग तो नहीं करने लगे हैं?

कहीं हम स्वयं ही अपने देश की प्रगति में रोड़ा तो नहीं बनते जा रहे?

दुनिया के समक्ष देश की तस्वीर को कहीं हमारे कुकृत्य ही तो बदरंग नहीं कर रहे?

कहीं हमसे जाने-अनजाने में ऐसा कुछ तो लिखा-कहा नहीं जा रहा जिस पर हमारी पीढ़ियाँ भी शर्मिंदा हों!

देश की उन्नति के स्वप्न को मन में सँजोए कहीं हम अपने कर्तव्य-पथ से भटक तो नहीं रहे हैं!

यदि इन सब का उत्तर 'न' है तो हमें न केवल हमारी

स्वतंत्रता के मूल्यों का ज्ञान है बल्कि हम दूसरों का भी मार्गदर्शन कर सकते हैं। करना ही होगा! इस पावन मातृभूमि के अनगिनत वीरों के त्याग और बलिदान से मिली स्वतंत्रता अमूल्य है, इसे हम किसी भी नासमझी से जाया नहीं होने दे सकते हैं।

आइये हम सब प्रण लें कि हमें स्वतंत्रता मिलेः-

उन विचारों से जो हमारी भावनाओं को हिंसक बनाने की कुचेष्टा करते हैं।

अधिकारों की उस माँग से, जो अपने कर्तव्य भुला देश की संपत्ति को नष्ट करने पर आमादा है।

उस सोच से जो हमारी संस्कृति और सभ्यता को अपमानित करने से नहीं चूकती।

उस अंधभक्ति से जहाँ धर्म की आड़ में हजारों युवाओं के मन मस्तिष्क में जहर भर दिया जाता है।

उस भेदभाव से जहाँ हजारों निर्धन और विवश इंसानों की मृत्यु से किसी को अंतर नहीं पड़ता, उनकी चर्चा नहीं होती। उन्हें उनके हिस्से का सम्मान तक नहीं मिलता। उनके जीवन का मोल बस एक निश्चित धनराशि तय कर दिया गया है।

जहाँ पैसा ही रिश्तों की धुरी बन जाता है और भावनाओं की कद्र नहीं होती।

जहाँ स्वार्थसिद्धि के लिए झूठ और मनगढ़ंत आरोप का सहारा ले किसी को मृत्यु द्वार तक पहुंचा नित झूटी कहानियाँ गढ़ी जाती हैं।

जहाँ स्त्रियों के प्रति हुए अपराध में दोषी को पकड़ने के पहले स्त्रियों की ही गलती ढूंढी जाती है। उनके वस्त्रों की चर्चा होती है। चरित्र हनन किया जाता है।

अभिव्यक्ति की स्वतंत्रता के नाम पर देश को ही दुनिया के सामने उछाला जाता है।

जहाँ प्रेम से कहीं अधिक घृणा, ईर्ष्या और वैमनस्यता के भावों को खाद पानी दिया जाता है।

जहाँ बच्चों के कोमल हृदय में भय भर दिया गया है। उनके बचपन पर तमाम बोझ लाद दिया जाता है और उन्हें खिलने नहीं दिया जाता।

राजनीति के उस छिछले दरिया से जिसके लिए जनता एक प्यादा भर है।

उन नेताओं से जो अपने लिए सम्मान की अपेक्षा रखते हुए नित दूसरों को अपमानित करने का एक भी मौका नहीं छोड़ते।

उस मानसिकता से जिसने हमारी आंखों पर पट्टी बांध नेताओं को हमारा खुदा बना दिया।

समाज को भ्रमजाल में लपेटने के लिए बने उन समस्त फर्जी

एकाउंट्स से जो झूठ के पहाड़ पर खड़े हो सच का थोथा आह्वान करते हैं।

गर्व से कहो कि हम भारतीय हैं:-

हम मानते हैं कि प्रत्येक नागरिक को शिक्षा चाहिए, रोजगार चाहिए और सम्मान से जीने का हक भी। लेकिन हमें उस दिन की भी प्रतीक्षा है, जब हमारे भारत का निर्धन, अशिक्षित, शोषित, पीड़ित और बेरोजगार वर्ग भी पूरी शक्ति के साथ स्वयं की उन्नति हेतु हर संभव प्रयास करे। इस प्रक्रिया में आरोप-प्रत्यारोप की कीचड़ से अपने-आप को दूर रख न केवल सरकार बल्कि प्रत्येक भारतवासी उनके साथ हो और हम सब हृदय से यह भाव महसूस कर पूरे जोशोल्लास के साथ सम्पूर्ण विश्व के सामने यह नारा गुंजायमान कर दें कि 'भारत हमको जान से प्यारा है, सबसे प्यारा हिन्दोस्तां हमारा है!'

इस जन्मभूमि पर सौ-सौ जीवन कुर्बान!

कहाँ है, काबुलीवाला?

अफगानिस्तान से अमेरिका की विदाई की आखिरी तारीख, 31 अगस्त 2021 थी। अमेरिकी सैनिकों ने तय तिथि से पहले अफगान ज़मीन तो छोड़ दी लेकिन साथ ही अरबों मूल्य के हथियार भी वहीं छोड़ दिए। इनमें से कुछ निष्क्रिय अवश्य कर दिए गए लेकिन फिर भी इनकी संख्या मानवता को थर्राने के लिए काफी है। इन हथियारों एवं अन्य खर्च हुई कीमत का हिसाब लगाना आम इंसानों के बस की बात भले न हो पर इनसे हुए विध्वंस के साक्षी हम सब हैं। एयरपोर्ट पर हुए धमाके भी सबने सुने। विचारणीय है कि क्या बदले की आग में धधकते अमेरिका ने पाने से कहीं अधिक खो दिया है? क्या हम यह मान लें कि अमेरिका ने वाकई बदला ले लिया? क्या वर्षों से चले आ रहे इस मसले के परिणाम सुखद दिखाई दे रहे हैं?

जो भी हो, लेकिन अगस्त 2021 का महीना अफगानिस्तान में इंसानी क्रूरता की महापदा के रूप में याद किया जाएगा। आगे क्या होने वाला है, उसकी कल्पना करते ही दिल दहल उठता है। उस पर विडंबना यह कि तालिबानियों के अन्याय के विरुद्ध चीखने वालों के सुर एन मौके पर बदल गए हैं। महाशक्तियाँ, तालिबान से दुश्मनी लेने के लायक हैं ही नहीं! तभी तो अपने घर में तमाम बुरी-भली बातें कहने के बाद भी सरकार के रूप में वे संभल कर बोल रहे हैं। इस समय शांतिप्रिय बने रहना उनकी विवशता है। वे विवश हैं इस निर्मम और क्रूर अध्याय को नजरअंदाज करने के लिए। तालिबान के आगे दुनिया की चुप्पी हमें ये समझा रही है कि ताकत और स्वार्थ के आगे भावनाओं और संवेदनाओं की बातें करना कितना निरर्थक है। देखिए, विज्ञान और तकनीक पर उछलती दुनिया ने अपना क्या हाल कर डाला है!

मूर्खों के हाथ में जब सत्ता आती है तो उनका व्यवहार कैसा होता है, ये जानना हो तो काबुल के मनोरंजन उद्यान एवं राष्ट्रपति भवन में घुसे तालिबानियों के चेहरे और हरकतें याद कीजिए। अफगानिस्तान पर तालिबान का कब्जा मात्र एक देश की ऐतिहासिक घटना नहीं है, इसने दुनिया का कच्चा चिट्ठा भी खोलकर रख दिया है। अत्याचार के विरुद्ध डटकर खड़े होने और मानवता का साथ देने वाली तमाम बातों, उपदेशों की धज्जियाँ उधेड़कर रख दी हैं।

हमें तो टैगोर का वो काबुलीवाला याद है जो बड़ा रहमदिल था। बच्चों को प्यार करता था। जिसने मिनी की झोली काजू, बादाम, किशमिश से भर दी थी। उसके हृदय में संवेदना थी, प्रेम था। बच्चों के लिए दुआएं थीं।

आज तालिबानी लाशों के ऊपर से चलते हुए काबुल पहुँचते हैं। यहाँ आने के बाद अपने सुधरने का ढोंग रचाए फिर रहे हैं कि 'देखो, हम स्त्रियों को ये करने देंगे, वो करने देंगे'। लेकिन बीते दिनों में वहाँ स्त्रियों की क्या स्थिति है, ये किसी सी छिपी नहीं! महाशय, आप ये भी तो बताइए कि उन स्त्रियों से क्या छीन लिया गया है। क्या शेष है अब? कोई अपना जीवन कैसे जिए, ये तय करने वाले आप होते कौन हैं?

एक खूबसूरत देश की चहकती तस्वीर पर दुख और आतंक की काली स्याही कैसे पोत दी जाती है, इस निराशा और अपार भय के साये तले, अफगानिस्तान के निवासी जी रहे हैं। जीना तो खैर क्या ही कहें! भूखे, बेघर, रोते-बिलखते मर ही रहे हैं। निर्दोष, निरपराध बच्चे, पुरुष-स्त्रियाँ सब अमानुषिक व्यवहार के शिकार हैं। एयरपोर्ट की स्थिति हमने देखी ही है! हवाई अड्डा युद्ध भूमि की जगह ले चुका था। वहाँ चेक पोस्ट बन गए थे। सेनाएं तय कर रहीं थीं कि कौन जाएगा!

भगदड़ और जान बचाने की आपधापी इस हद तक भयावह रही कि लोग प्लेन को बस की तरह पकड़ रहे थे। उसके ऊपर जहाँ-तहाँ लटक जीवन छूटता रहा है। हर तरफ गोलियों की आवाज, चीख पुकार के बीच एक दीवार पर नजर जाती है। एक बच्चा हवा में झूल रहा है। उसके लिए इस दीवार को फाँदना जीवन या मृत्यु के द्वार का माध्यम प्रतीत नजर आता है। ये बच्चा और इसके जैसे तमाम बच्चों का भविष्य क्या होगा! स्त्रियों के दर्द और गुलामी की नई इबारत कितने खौफनाक तरीके से लिखी जाएगी! पुरुषों के एक इन्कार से कितनी क्रूरता से उनका सिर कलम कर दिया जाएगा! हम इन्हें गिन नहीं पाएंगे। बस, शर्मिंदगी से सिर झुका रहेगा।

इनकी दर्दनाक कहानियाँ इतिहास में लिख दी जाएंगीं। साथ ही यह भी लिखा जाएगा कि एक शासक के भाग जाने से देश की ज़मीन कैसे खून से रंग जाती है। देशवासियों की चीखें कैसे आसमां को स्तब्ध कर देती हैं! ये निर्मम आतंकियों के काबुल पर कब्जे का किस्सा तो है ही पर एक राष्ट्रपति की अपने देशवासियों को उनके हाल पर छोड़ भाग जाने की शर्मनाक घटना भी है। ये इतिहास को शर्मसार कर देने वाले धोखे का वह स्याह किस्सा है। जिसे बची हुई प्रजा आने वाली नस्लों को सुनाएगी।

यह घटना, इस पृथ्वी के हर निवासी को अपना नेता चुनने से पहले हजार बार सोचने पर विवश कर देगी। हम सब इस भीषण दुखद मंजर के साक्षी बन रहे हैं कि शासक के भाग जाने से देश की कैसी दुर्गति होती है। इसने जनता और शासक के बीच अविश्वास की गहरी खाई खोद दी है। आज अफगानिस्तान को उसके आकाओं

सहित सारी दुनिया ने उसके हाल पर छोड़ दिया है।

लेकिन जिसका कोई नहीं होता, अब तो खुदा भी उसका साथ देता नहीं दिखता! दुनिया से क्या और कैसी उम्मीदें! वह तो सदा ही ऐसी रही है। स्वार्थी, लालची और कभी मजबूर भी।

पैसे ने सदा ही मानवता को हराया है। घर, परिवार, समाज कहीं भी देख लीजिए, पैसा ऐसे ही मुँह बंद कर देता है। जब अपनी पर आती है, तो दिन-रात सुविचार बाँटने वाले लोग ऐसे ही मौन धारण कर लेते हैं। तालिबानियों के अत्याचारी व्यवहार पर अधिकांश देशों की चुप्पी, इसी का ताजातरीन उदाहरण है। यकीनन यह आतंकवाद का समर्थन भी नहीं, लेकिन पूंजीवाद की विवशता है। दुनिया का कोई भी देश हो, वह स्वयं की आर्थिक हानि नहीं होने देना चाहता!

हम भी व्यक्तिगत तौर पर अन्याय और अधर्म के विरोध में जो जी चाहे बोल लें लेकिन अंततः मानना वही होगा जो देश हित में है। सरकार ही अन्य देशों से रिश्ते तय करती है और उस स्थिति में जो भी निर्णय हो, हमें स्वीकारना होगा। यह भी नहीं भूलना चाहिए कि हमारे कई नागरिक अभी भी काबुल में हैं और उनकी सुरक्षा सर्वोपरि है। हमारे हाथ में बस इतना है कि हम अपने बच्चों को नफरत न सीखने दें। उन्हें प्रेम और मानवता का पाठ इस उम्मीद से पढ़ाते रहें कि आने वाली पीढ़ियाँ शांतिप्रिय, न्यायप्रिय और हथियारों से दूर रहने वाली बनें। हम तमाम अफगानियों के लिए प्रार्थना और उनकी कुशलता की कामना करें।

ओ, प्रिय काबुलीवाले! मिनी और अफगानिस्तान के सारे बच्चे तुम्हें बहुत याद कर रहे हैं। तुम्हारे देश के बच्चों की आँखों की चमक खो चुकी है। उनके चेहरे का नूर चला गया है। इनकी झोली में सूखे मेवे भरने कोई नहीं आया!। आज इस दुनिया को, मिनी को, हम सबको एक और काबुलीवाले रहमत की दरकार है!

अफगानी स्त्री की पाँच तस्वीरें

जिस दिन हम अपना स्वतंत्रता दिवस मनाने में मशगूल थे, ठीक उसी दिन तालिबान, अफगानी महिलाओं की गुलामी का नया अध्याय रच रहा था। अपने क्रूर इरादों का महल खड़ा करने का स्वप्न लिए आतंकियों की पलटन जब काबुल पहुँची तो वहाँ कोहराम मच गया। वहाँ से आती तस्वीरों ने सबकी आत्मा को झकझोरकर रख दिया है। अब खबर ये है कि इन दिनों वहाँ बुर्के की कीमत और बिक्री दोनों बढ़ गई है। लेकिन मानवता किस रसातल में चली गई है उसे नापने का कोई पैमाना नहीं है।

तालिबान के काले इतिहास पर मैं बात नहीं करूंगी। मैं अफगानिस्तान पर उनके कब्जे के कारणों की विवेचना भी नहीं करूँगी, उसके लिए जानकार लोग बैठे हैं। मैं धर्म और राजनीति की भी कोई बात नहीं करना चाहती। लेकिन एक प्रश्न है जो लगातार साल रहा है कि हर मुसीबत की गाज महिलाओं पर ही क्यों गिरती है? देश, चेहरे, नाम बदलते हैं, लेकिन महिलाओं के साथ होने वाला सुलूक नहीं उनके शोषण और यातना की कहानियाँ थमने का नाम ही नहीं लेतीं।

अफगान महिलाओं पर जो पाबंदियां, सख्त कानून लादे जा रहे हैं, उसमें उनका क्या दोष है। न तो वो अमेरिका को अफगानिस्तान बुलाकर लाई थीं। और न ही उन्होंने तालिबान को आने से रोका। खैर, अब तालिबान से मुखातिब और कोई बात नहीं होगी। मैं तो दुनिया की तरफ देखना चाहती हूं। हमारे, आपके जैसे मानवतावादी लोगों की तरफ। क्या ये तस्वीरें आपका दिल नहीं झकझोरती?

पहली तस्वीर: जान आफत में डाल दुधमुंहे बच्चे को नाटो सैनिक के सुपुर्द करती बेचारी अफगान मां

कितनी पीड़ा और वेदना छुपी है, इस एक तस्वीर में। कैसा अकल्पनीय और रोंगटे खड़े कर देने वाला दृश्य है यह। सोचिए, जिस दृश्य को देख हमारा कलेजा हिल गया, तो उस महिला पर जाने क्या-क्या न गुजरी होगी! आखिर कुछ तो मजबूरी रही होगी कि एक मां घर से निकल, अपने दुधमुंहे बच्चे को लेकर काबुल एयरपोर्ट पर आ खड़ी हुई। इतने मर्दों के बीच तालिबानियों के जबड़े से बचकर वो अपनी, अपने बच्चे की जान जोखिम में डाल क्यों आई होगी? वजह साफ है, उसको पता है कि यदि इस एयरपोर्ट से उसे हवाई जहाज मिल गया, तो जिंदगी की सुबह हो जाएगी। यदि न जा सकी, तो यहां होने वाली जिंदगी का अंधेरा मौत से बदतर होगा। उसके लिए भी और उसके बच्चे के भविष्य के लिए भी।

कंटीले तारों से पटी दीवार के इस पार आना उसकी आखिरी उम्मीद के जैसा है। और उसकी ये छटपटाहट सारी दुनिया को दिखती है। आने वाले इन बच्चों का क्या भविष्य होगा? उनके परिवार तक वो कैसे पहुंचेंगे? कभी पहुंचेंगे भी या नहीं? उन बच्चों का क्या होगा, जिन्होंने अपनी आँखों के सामने मौत का नंगा नाच देखा है! क्या दुनिया का कोई भी मनोवैज्ञानिक, उनके कोमल मन पर हुए इस वीभत्स प्रहार को, उसके दुख को कम करवा पाएगा? क्या ये कभी सामान्य बच्चों की तरह मुस्कुरा पाएंगे? अनगिनत बच्चों के चेहरे भीतर तक उतर मन को बार-बार कचोटने लगते हैं। उनकी आँखों में ये प्रश्न बार-बार नजर आता है कि उन्हें इस बदसूरत दुनिया में किसलिए लाया गया है? किसकी तरस के सहारे उनकी किस्मत में जिंदा रह पाना बदा है? ये प्रश्न शर्मिंदा करते हैं लेकिन इनके उत्तर ढूंढे नहीं मिलते!

दूसरी तस्वीरः एयरपोर्ट के दरवाजे पर गुहार लगाती बिलखती लड़कियां

आपने एयरपोर्ट के दरवाजे पर हेल्प, हेल्प कहकर सैनिकों को पुकारती, जीवन-रक्षा की गुहार लगाती, रोती-बिलखती लड़कियां भी जरूर देखी होंगी। आखिर वो भयाक्रांत लड़कियाँ क्या माँग रही हैं? जीने का अधिकार ही न! जब हम ऐसी तस्वीरें देखते हैं तो मानवता के सबसे निकृष्ट दौर में होने का दुख सालने लगता है। जीवन का ऐसा निरीह रूप भी कभी सामने आ खड़ा होगा, ये कहाँ जानते थे हम!

वो रेलिंग जिसके इस पार आने के लिए ये चीत्कारें भर रहीं हैं तो केवल आना भर ही उनका मकसद नहीं है। वे परी कथाओं के स्वप्नलोक की इच्छा नहीं कर रहीं। वे तो गुलामी की तमाम बेड़ियों को पार करना चाहती हैं। नारकीय जीवन से मुक्ति की आस लगाए बैठी हैं। एक सामान्य मनुष्य की तरह अपना जीवन जीना चाहती हैं। क्या एक देश की महिलाओं की ये मांग बहुत ज्यादा और नाजायज है? उनकी इस हृदयविदारक पुकार पर दुनिया भर की महाशक्तियों की चुप्पी असहनीय, अक्षम्य है। अभी तो भीतर की कुछ खबरें बाहर आ भी रही हैं। बाद में जो होगा उसकी रत्ती भर भी भनक तक न लगेगी किसी को।

तीसरी तस्वीरः यदि 20 दिन की मोहलत है तो ये जिंदगी आजादी के नाम होगी

भय और आतंक की तमाम खबरों के बीच एक और दृश्य सामने से गुजरा। एक चेहरा अन्याय के विरुद्ध सामने डटकर खड़ा था। यह लड़की पूरी जीवटता के साथ महिलाओं पर हुए इस अत्याचार का विरोध कर रही है। यह विरोध प्रदर्शन वाले उस जुलूस का हिस्सा थी, जहाँ दौ सौ के हुजूम में सात महिलाएं भी उपस्थित

थीं। यह कहती है, 'पिछले 19 वर्षों से मैं पढ़ाई कर रही हूं और अपने लक्ष्यों को हासिल करने के लिए प्रयास कर रही हूं लेकिन दुर्भाग्य से आज मेरे सारे सपने मर गए।'

उसने बताया कि 'एक तालिब ने कहा, मैं जो भी करना चाहती हैं, उसके लिए सिर्फ 20 दिन की ही आजादी है।' इससे ये अनुमान सहज ही लग जाता है कि उसके बाद कुछ भी हो सकता है।

वो दृढ़ता से कहती है, 'मैं इन 20 दिनों को अपनी आवाज बुलंद करने के लिए इस्तेमाल करना चाहती हूं। हर कोई तालिबान से डरा हुआ है लेकिन मैं नहीं जब तक वो लोग मुझे शूट नहीं कर देते, मैं अपने मौलिक अधिकारों के लिए लड़ती रहूंगी। उन्हें वो मुझसे छीन नहीं सकते।'

ये तय है कि भीगी आँखों वाली यह लड़की, 20 दिनो में अपनी जी जान लगा देगी। वो अपने सपनों और बीते वर्षों की पढ़ाई को आसानी से जाया नहीं होने देगी। इसकी आवाज देश के झंडे पर गर्व करने वाली और उसे अपनी पहचान कहने वाली लाखों महिलाओं की आवाज है। इस आवाज का दुख हर अफगानी लड़की की प्रतिध्वनि है। दुनिया भर की सताई महिलाओं की पीड़ा इसके चेहरे पर झलकती है। इसके हौसलों को सलाम। सच है, जब मौत तय है तो बेशक काल से जूझते हुए ही क्यों न हो!

दुआ कीजिए, कि हर महिला की आँखों में यही जुनून दिखाई दे। आखिर इस दर्द को कभी तो विराम मिले। कोई तो पल आए, जब ये अपने सपनों में अपनी ख्वाहिशों के रंग भर सकें। एक तरफ इनकी जिजीविषा है तो दूसरी तरफ घटाटोप अंधेरा छाया हुआ है। इस काले अंधकार का भय तो गले से एक निवाला तक नीचे नहीं उतरने देता। मानवता के दुश्मनों का हमेशा के लिए नाश क्यों नहीं हो जाता!

चौथी तस्वीरः अज्ञातवास में जीती आतंक और यातनाओं की दास्तानें

गुलामी में जकड़ी महिलाएं, मौत से बदतर जिंदगी जीने को मजबूर कर दी जाती हैं। अफगानिस्तान से कुछ वर्ष पहले भारत आई एक महिला बताती हैं कि उनकी दो बेटियां थीं जिनमें से एक की 12 वर्ष की उम्र में ही जबरन शादी करा दी गई। दूसरी को कम उम्र में ही बेच दिया गया। वे दोनों ही भीषण शारीरिक एवं मानसिक यंत्रणा से गुजरती रहीं। उसके बाद उनकी खबरें मिलना बंद हो गईं। ये जानकर रूह कांप जाती है कि उन दोनों को इस दारुण जीवन में धकेलने वाला जल्लाद उनका पिता ही था। बेटियों के गुजर जाने के बाद ये महिला जैसे-तैसे जान बचाकर यहाँ आई और अब शिक्षा प्राप्त कर जीने की कोशिश कर रही है।

ये किस्सा तो एक बानगी भर है। आदिमकालीन रीति रिवाजों वाले क्रूर, यातनापूर्ण शासन में महिलायें आत्मनिर्भर न होकर, बंधुआ मजदूर बना दी जाती हैं। दोषियों पर कोड़े और पत्थर बरसाए जाते हैं। महिलाओं को घर से बाहर काम नहीं करने दिया जाता। हिजाब पहनकर निकलना होता है। 12 वर्ष से ऊपर का होते ही बच्चियों से शिक्षा का अधिकार छीन लिया जाता है। उस पर दिन-रात आतंक के साये में जीना कितना खौफनाक है, उसका तो क्या ही कहा जाए!

पांचवी तस्वीर: जो दिखाई नहीं देगी, दरिंदे उन्हें नामालूम मौत देंगे।

प्रताड़ना की शिकार असहाय महिलाओं में से कुछ चेहरों के नाम होंगे, कुछ की पुकार हम तक पहुँच सकेगी लेकिन उन महिलाओं की पीड़ा का क्या, जिनकी तस्वीर नहीं है, और न ही आएगी। लेकिन तालिबान शासन में वो रोज नई यातना की शिकार बनेंगी। कम उम्र में उनकी जबरन शादी होगी या उन्हें बेच दिया जाएगा। उनका बलात्कार होगा और फिर निर्मम हत्या भी। इनकी चीखें जबरन थोपे गए लिबास के भीतर ही टकराकर दम तोड़ देंगी। इनकी घुटन के दर्द का गवाह कोई न रहेगा। इनका वजूद और नाम दोनों मिटा दिए जाएंगे। इन लाशों की संख्या इतिहास के पन्नों में दर्ज भी न होगी कभी।

अमेरिकी बलों ने शनिवार तक अफगानिस्तान से लगभग 12,700 लोगों को निकाला है, राष्ट्रपति जो बाइडेन ने अफगानिस्तान से सभी अमेरिकियों को अपने देश लाने का दृढ़-निश्चय कर लिया है। एक-एक कर सभी देश, अपने अपनों को निकाल लाएंगे लेकिन उन अफगानियों का क्या? उन अफगानी महिलाओं का क्या? क्या उन्हें जुल्म सहने और दोजख में तड़प-तड़पकर मरने को यूँ ही छोड़ दिया जाएगा? मनुष्य ही मनुष्य को खा रहा है। अब और क्या बचा है, देखने को!

हाँ, एक बात और, कि यूँ तो सारी बातें अफगानी महिलाओं के परिप्रेक्ष्य में ही की जा रहीं हैं लेकिन दुनिया भर के अन्य देशों में भी उन्हें कोई राजकुमारी बनाकर नहीं रखा गया है। उन पर हुए अत्याचार, यौन शोषण एवं मानसिक उत्पीड़न की घटनाएं वैश्विक स्तर पर दिनोंदिन बढ़ती जा रही हैं।

यदि आपका मन चाहे तो आप महिलाओं को मनुष्य प्रजाति के हिस्से के रूप में भी मत देखिए। बल्कि उन्हें एक ऐसी नस्ल ही मान लीजिये, जैसे विलुप्त होने वाला कोई जीव। शायद इसी बहाने उनका कुछ उद्धार हो जाए। शायद इसी बहाने ही सही, उन पर टूटती मुसीबतों का पहाड़ कुछ थम जाए और उनको इस 'सभ्यता' के इतिहास में जीवाश्म बनने से रोका जा सके!

मैं फिर कह रही हूँ कि अब ये धर्म और राजनीति का मसला है ही नहीं जीव दया का मामला है। दुनिया के आगे अब इससे ज्यादा कातर मनुहार और क्या करूं!

मिलावटी मनुष्य

मिलावट की कहानी कहाँ से प्रारंभ हुई, यह तो नहीं जानती लेकिन इतना अवश्य याद है कि पहले दूध में पानी मिलाने के किस्से आम हुआ करते थे। ये इतने आम थे कि बुरा ही नहीं लगता था। बड़ी ही सहजता से दूधवाले को कह दिया जाता कि 'आजकल पानी कुछ अधिक ही मिला रहे हो!' या कि 'दूध में मलाई ही नहीं पड़ती भैया, दही बिल्कुल अच्छा नहीं जमता!' वह सुनता और खिल्ल से हँस पड़ता। बात इसके आगे कभी बढ़ी ही नहीं!

फिर त्योहारों के समय मावे में मिलावट के समाचार मिलने लगे। हमने यह सोचकर मन को आराम दिया कि 'हम तो घर में ही बना लेते हैं'। मसालों और अनाज में भी यही होता रहा। स्त्रियाँ खड़े मसाले पीसतीं और अनाज साफ कर लिया करतीं। बीच-बीच में 'जहरीली शराब पीने से इतने लोगों की मृत्यु एवं इतने अस्पताल में भर्ती' के आँकड़े भी समाचार-पत्र में प्रमुखता से दिखाई देते। हम यह कहकर मुँह बिचका लेते कि 'इन शराबियों को तो वैसे ही मरना था!'।

कहने का तात्पर्य यह है कि जैसे प्रातःकाल को सूर्य उदय होता है, चिड़िया चहचहाती हैं, फूल खिलते हैं, धूप देहरी पर आकर पाँव पसार लेती है, चाय के सिप के साथ अखबार को पलटा जाता है, ठीक इतनी ही सहजता से 'मिलावट का होना' भी हमारी दिनचर्या में घुल चुका है। हमने कभी किसी से प्रश्न ही नहीं पूछा कि 'ऐसी स्थिति क्यों है? मिलावट करने वाले को क्या दंड दिया गया?' बल्कि इसके विपरीत भावुक होते हुए यह सोचा कि 'अरे, बेचारे गरीब लोग हैं! कमाने के लिए 'जरा सी बेईमानी' ही तो की है! इतना तो चलता है!' और ऐसा क्यों न मानते भला! हम स्वयं ही कौन से दूध के धुले थे! मनपसंद फिल्म लगी हो और टिकट खिड़की पर 'सोल्ड' का बोर्ड, तो हम तुरंत दोगुने दाम देकर ब्लैक में टिकट खरीद, स्वयं में विजेता वाली अनुभूति भर लेते। पद का रुतबा दिखाकर पंक्ति में आगे लग जाना भी कोई बड़ी बात नहीं थी। पुलिस वाले को या किसी बाबू को सौ-दो सौ पकड़ाकर तो न जाने कितने काम कराये हैं। लेकिन स्वयं को मन ही मन निर्दोष मानकर बरी भी करते रहे कि 'हम तो भ्रष्टाचार के विरोधी हैं, वो पैसे तो बस चाय-पानी को दिए थे'।

वैसे हम 'जुगाड़ तंत्र' के वरिष्ठ विद्यार्थी, इन सब परिस्थितियों से बचकर निकल आने में सक्षम हो चुके थे। यूं भी मध्यम वर्ग के लिए अपना काम करा लेना किसी सफलता से कम नहीं होता! निर्धन हमेशा पिसता रहा, वही मरता भी रहा। वो अशिक्षा का मारा

था और हम सभ्य, शिक्षित वर्गों के सरोकार से प्रायः दूर ही रहा।

धीरे-धीरे चाय-पानी के रेट बढ़ते रहे और मनुष्य की लालसा भी। अब सब्जी-फल को हानिकारक रसायनों से पकाने के समाचार मिलने लगे, हमने सुपरस्टोर से खरीदने के स्थान पर ठेले वाले या सीधा खेत से लाकर बेचने वालों से लेना शुरू कर दिया। मीडिया ने तमाम उत्पादों के असली-नकली पहचान के कार्यक्रम दिखाने प्रारम्भ कर दिए। हमारे बुद्धिमान मस्तिष्क ने तुरंत उन सबको आत्मसात कर लिया। लेकिन इन रसायनों ने धीमे-धीमे न जाने कितने मनुष्यों को मौत के घाट पहुँचा दिया होगा, उस तथ्य पर हम मौन ही रहे।

हमारे देखते-देखते ही, वस्तुओं की कालाबाजारी और नकली उत्पादों ने बाजार में गहरी पैठ बना ली, हमने तब भी कुछ नहीं कहा! हाँ, इस पर दुःख अवश्य जताने लगे थे।

समाज का यही रवैया, मानवता के दुश्मनों की हिम्मत बढ़ाता रहा और उन्हें भी इस बात का पूरा विश्वास हो गया कि तंत्र उनका कुछ नहीं बिगाड़ सकता! जमीर की कीमत, चंद रुपयों से सदैव ही चुकाई जाती रही है और रिश्वत की खेप कैसे ऊपर से नीचे तक बँटती आई है, ये भी कोई आश्चर्यजनक तथ्य नहीं रह गया है। वे नेता जो स्वयं नोट देकर, वोट खरीदते हैं, हम यदि उनके 'भ्रष्टाचार उन्मूलन' के नारे को सुन भावुकता में बह जाते हैं तो ये सरासर मूर्खता है। समाज में यदि कालाबाजारी बढ़ रही है तो ये न सोचिए कि मासूम राजनेताओं को इसकी खबर नहीं! वे तो स्वयं ही इस मिलावट और धोखे को प्रश्रय देते रहे हैं। वरना अपराधियों के हौसले इतने बुलंद नहीं होते!

अब कुछ राक्षसों की हिम्मत इतनी बढ़ गई है कि इन्होंने जीवन रक्षक दवाइयों और इंजेक्शन को भी नहीं बख्शा! एक इंजेक्शन जो 900 रुपए का आता है, अगर तीस हजार में मिल रहा है तो परिजन उसे खरीदने को विवश होंगे ही क्योंकि यहाँ किसी अपने के जीवन-मरण का प्रश्न है। इस समय हम भ्रष्टाचार की बात कर ही नहीं सकते! हमें कैसे भी और किसी भी मूल्य पर दवा चाहिए, इंजेक्शन चाहिए, ऑक्सीजन सिलेंडर चाहिए। यही मनोभाव बेड और अस्पताल के लिए भी है।

देखिए दूध की मिलावट से चलकर, आज हम इस हाल में पहुँच चुके हैं कि हमारे पास रोने-कलपने के सिवाय और कुछ भी शेष नहीं रहा! लालचियों ने निर्लज्जता की समस्त सीमाएं लांघ ली हैं। अब बस, इंसान का इंसान को काटकर खाना ही शेष है!

आज जो दुर्दशा है उसके लिए कोई एक व्यक्ति ही जिम्मेदार नहीं है। पूरा का पूरा समुच्चय है। जनता और सत्ता के चरित्र में भी

अधिक अंतर नहीं रह गया है। आपदा के समय या लाशों के साथ लूटपाट करने वाले हमारे ही बीच के लोग हैं। इनका नैतिक-अनैतिक कुछ नहीं है। ये एक बोतल शराब पाकर ही अपना नेता चुन लेते हैं। जो बड़े नाम हैं, उनकी व्यवस्था भी उनके पद के हिसाब से बड़ी हो जाती है। सुधार चाहा ही किसने?

यदि आप इस सिस्टम को बदलना चाहते हैं तो मूकदर्शक न बने रहें। चेत जाइए और प्रश्न करना न भूलिए! अव्यवस्था के विरुद्ध चीख कर बोलिए! जो इस काली व्यवस्था के अपराधी हैं और जिनके वरदहस्त में ये सब पल रहा है उनकी सूची तैयार कीजिए और उनके लिए फांसी से कम की मांग न कीजिए।

ये कभी न भूलिएगा कि अगर 'चलता है' को चलाते रहे तो एक दिन, हम सब भी अचानक चले जाएंगे और हमेशा-हमेशा के लिए भुला दिए जाएंगे। आखिर, मनुष्य की संवेदनाएं भी कम मिलावटी थोड़े न हैं!

क्या है, भारतीयों की जान की कीमत?

क्या स्वयं को जीवित बनाए रखने के लिए, अपनी भूमिका शिद्दत से निभाई है हमने?

जिस तरह जन्म एक शाश्वत सत्य है, वैसे ही मृत्यु भी! जन्म लेते ही पहली साँस के साथ हमें एक नाम मिलता है, पहचान मिलती है। और जैसे ही साँसों की ये डोर थमती है, वह पहचान खो जाती है। हमारा नाम अब किसी शव में बदलकर अपना अस्तित्व खो चुका होता है। वह शरीर जिसे सर से लेकर पाँव तक सजाने में हमने एक उम्र बिता दी, वह किसी सजावट का मोहताज नहीं होता! उसे मौसमों का फ़र्क़ नहीं पड़ता! लेकिन मनुष्य की प्रवृत्ति ऐसी है कि जब तक साँसें हैं, तब तक वह जीवन को बचाए रखने का संघर्ष जारी रखता है। तमाम सपने साथ लेकर जीता है। दुख यह है कि जिस अभिशप्त दौर में जी रहे हैं, वहाँ ये जीवन ही सबसे आसानी से खो जाने वाली शय है।

अब यह तथ्य तो जगजाहिर है कि कोरोना काल में गंगा में बहती हुई, या अपने अंतिम संस्कार की प्रतीक्षा ढोती हुई लाशें यहाँ-वहाँ पड़ी मिल रही थीं। लेकिन बात केवल महामारी के इस दौर से ही जुड़ी नहीं है, बीते समय को खंगालकर देखें तो ऐसे सैकड़ों चेहरे सामने आ जाएंगे। क्या हमने कभी इस विषय पर विचार किया है कि आखिर हम भारतीयों की जान की कीमत क्या है?

व्यक्तिगत तौर पर हम स्वयं की कितनी परवाह करते हैं?

कहते हैं मानव जन्म हजार योनियों के बाद मिलता है। संभवतः इसलिए ही इसकी महत्ता सर्वाधिक है। यूं भी जीवन सभी को प्यारा होता है। हर कोई चाहता है कि वह शतायु हो। मनुष्य इसके लिए तरह-तरह के जतन करता है। स्वस्थ, संतुलित खानपान, ध्यान, योगाभ्यास, मल्टी-विटामिन का उपयोग, समय-समय पर अपने स्वास्थ्य की जाँच इत्यादि हमारी इसी जद्दोजहद का हिस्सा हैं। लेकिन इन्हीं के बीच में एक वर्ग ऐसा भी है जो जान हथेली पर लिए घूमता है।

'कोरोना काल में मास्क न पहनने वाले और वैक्सीन न लगवाने को लेकर कुतर्क करने वालों की भी कोई कमी नहीं! ये न केवल अपनी जान को जोखिम में डाल रहे हैं बल्कि अपने आसपास के लोगों को संक्रमित करने का भय भी उत्पन्न कर रहे हैं।

हेलमेट न पहनने वाले इसमें सबसे ऊपर हैं। दुपहिया वाहनों की सड़क दुर्घटनाओं में अधिकांश ऐसी होती हैं कि यदि वाहन चालक ने हेलमेट पहना होता, तो उसकी मृत्यु नहीं होती! कुछ लोग गर्मी के कारण इसे न लगाने का बहाना बनाते हैं, तो किसी को

अपने केश विन्यास के बिगड़ने का भय सताता है। कोई बस इसलिए नहीं लगाता कि यहीं पास में ही तो जाना है। लेकिन दुर्घटना कभी कहकर या आपके गंतव्य को देखकर नहीं होती!

सड़कों पर सर्पाकार अंदाज में बाइक लहराते, तमाम कलाबाजियाँ दिखाते युवाओं की तस्वीरें आए दिन खबरों में दिखाई देती हैं।

खतरनाक तरीके से सेल्फी लेने के कारण अपनी जान गंवा बैठने की घटनाओं की भी कोई कमी नहीं! यहाँ जान की कीमत सोशल मीडिया पर मिले लाइक की संख्या के सामने बौनी आंक ली जाती है।

प्रेम, नौकरी, व्यापार, परीक्षा में असफलता के कारण आत्महत्या करने वालों की भी बहुत बड़ी संख्या है। यहाँ रिश्ता टूटने, उम्मीद के अनुसार पैसा न कमा पाने, उधारी न चुका पाने या किसी की उम्मीदों पर खरे न उतर पाने का दुख इतना गहरा हो जाता है कि मनुष्य घनघोर निराशा और अवसाद में डूब, अपना जीवन समाप्त करने को ही समस्या का समाधान मान लेता है। वह यह भूल जाता है कि जीवन संघर्ष का ही दूसरा नाम है। वह यह भी नहीं सोच पाता कि उसके जाने के बाद उस पर आश्रित लोगों का क्या होगा!

इन सारे तथ्यों के अलावा आम आदमी का अपने स्वास्थ्य के प्रति लापरवाह होना भी बहुधा जानलेवा साबित होता है। किसी भी बीमारी के प्रारंभिक लक्षणों को हम प्रायः गंभीरता से नहीं लेते! हम यह मानकर चलते हैं कि हमें तो कुछ हो ही नहीं सकता! कुछ ऐसे भी लोग हैं जो देसी उपाय या टोने टोटके से खुद ही इलाज करने लगते हैं। होश तब आता है, जब बीमारी जब पूरे शरीर में अपने पाँव पसार लेती है। लेकिन उस समय सिवाय अफसोस के कुछ हाथ नहीं बचता!

समाज के लिए हमारी कीमत क्या है?

स्कूल के दिनों में पढ़ी गई पुस्तकों ने समझाया कि समाज, ऐसे लोगों का समूह है जो मानव जाति की स्थिरता के लिए परस्पर मिल कर रहते हैं तथा सामाजिक कल्याण के उद्देश्यों की पूर्ति का समर्थन अथवा उस हेतु कार्य करते हैं। यद्यपि कोरोना काल में व्यक्तिगत तौर पर एवं कई समाजसेवी संस्थाओं के सराहनीय कार्यों को देखते हुए इस परिभाषा को पूरी तरह नकारा नहीं जा सकता परंतु देश-विदेश की बदलती हुई परिस्थितियों पर दृष्टि डालें तो तमाम विध्वंसकारी गतिविधियों को देखते हुए, यह परिभाषा अत्यंत खोखली ही नजर आती है।

'समाज के लिए किसी की मृत्यु से उपजा दुख इस बात पर निर्भर करता है कि उस व्यक्ति विशेष से उसके रिश्ते कैसे थे! व्यक्तिगत तौर पर इस क्षति को मृतक के माता-पिता, पति/पत्नी, बच्चे, भाई बहिन, प्रेमी/प्रेमिका या करीबी मित्र के अतिरिक्त और कोई महसूस नहीं करता! शेष सभी की संवेदनाएं सामाजिक औपचारिकता भर हैं जो नमन और विनम्र श्रद्धांजलि लिखते ही समाप्त हो जाती हैं। देखा जाए तो कुल जमा चार लोग ही निकलेंगे, जिन्हें आपकी मृत्यु से दुख होगा।

यदि किसी बड़ी संस्था/कंपनी के प्रमुख का निधन हो गया हो तो सबको उस दुख से कहीं अधिक अपनी नौकरी की चिंता सताती है।

कोई, किसी का उधार चुकाए बिना परलोक सिधार गया हो तो उस पैसे के डूबने के अफसोस में आँसू ज्यादा निकलते हैं।

किसी प्रिय प्रसिद्ध व्यक्ति के देहावसान का दुख अवश्य होता है पर उससे निजी जीवन तनिक भी प्रभावित नहीं होता।

कुल मिलाकर समाज के लिए हमारी कीमत, एक वस्तु की तरह है, जिसकी उपयोगिता उसकी उपलब्धता और निहित स्वार्थ से तय होती है। जीते जी जिसे पलटकर भी नहीं देखते, मरने के बाद उसे अपनी आँख का तारा बनाकर पेश करना, समाज के दोगलेपन का सबसे अश्लील रूप है।

सरकार के लिए हम कितने महत्वपूर्ण हैं?

सरकार के लिए हमारी कीमत, एक वोट के बराबर है, जिसे कभी भी औने-पौने दामों में खरीदा जा सकता है। चाहे किसी की भी सरकार रही हो, जनता उनके लिए उपभोग किये जाने वाला जीव ही रही है। जिसकी आवश्यकता, केवल चुनाव के समय महसूस होती है। यहाँ तमाम प्रलोभनों से बरगलाकर हमारी संवेदनाओं से खेला जाता है और उसके बाद कोई सुध लेने वाला नहीं होता।

हम संख्याओं में गिने जाते हैं। धर्म के तराजू पर तौले जाते हैं।

सरकारें शिक्षा और चिकित्सा व्यवस्था को दरकिनार कर, हमारी धार्मिक भावनाओं को भुना अपना उल्लू सीधा करती रहीं। इस प्रक्रिया में आप मरें या जियें, किसी को फर्क नहीं पड़ता! मुआवजे की घोषणा अवश्य हो जाती है, जिसे पाने में परिजनों का ये जन्म तो निकल ही जाता है।

महामारी के इस समय में, इलाज के लिए भटकते उन लाखों लोगों से पूछिए कि उनको अस्पताल पहुँचाने, दवाई उपलब्ध कराने में किसका योगदान अधिक रहा?

सरकारें चाहती रही हैं कि हम किसी गुलाम की तरह उनकी

वाहवाही करते रहें लेकिन कभी, कोई प्रश्न करने का दुस्साहस न करें! उनकी नजरों में अपनी कीमत का अनुमान, हमें स्वतः ही हो जाना चाहिए।

सरकार के लिए केवल और केवल चाटुकारों की महत्ता है। हालाँकि उनकी मृत्यु से भी वह विचलित नहीं होती। शेष सभी मात्र आँकड़ा भर हैं जिसके कम हो जाने से उन्हें कोई फ़र्क़ नहीं पड़ता, बस छवि चमचमाती रहे!

जरूरत पड़ने पर, जब-जब सरकार ने देशवासियों को कहा, सबने खुलकर साथ दिया। दिल से सहायता भी करते हैं। लेकिन जब हमारी सहायता का समय आता है तो हमें तिल-तिलकर मरने को छोड़ दिया जाता है और फिर केंद्र तथा राज्य सरकारों द्वारा, मौत का ठीकरा एक-दूसरे पर फोड़ने का तमाशा जारी रहता है। इस सब में हम कौन से पायदान पर हैं, और नेताओं की रुचि किसमें ज्यादा है, खुद ही समझ जाइए।

वे सरकारी नेता, जिनके दर्शन भर से आम आदमी धन्य महसूस करता है और उनके कदमों में बिछा जाता है। एक दिन वही उसे आसानी से कुचल आगे बढ़ जाते हैं।

साफ बात यह है कि हमें अपने-आपको स्वयं ही सुरक्षित रखना होगा क्योंकि सरकार के लिए आम आदमी की कीमत सदैव ही दो कौड़ी की रही है!

विडंबना यह है कि भय या मजबूरी से, हम बहुत सी चीजों के आदी होते जा रहे हैं। इसके अलावा और कोई सरल विकल्प भी नहीं नजर आता! जान सबको प्यारी है। हमारा सारा जीवन इसे बचाने में ही बीत जाता है और फिर एक दिन किसी आपदा के चलते अनायास ही इससे भी हाथ धो बैठते हैं।

लेकिन क्या हमारी कीमत क्या सचमुच इतनी कम है? क्या हमें सबसे पहले इसे तय करने पर विचार नहीं करना चाहिए? आत्मरक्षा के लिए नहीं सोचना चाहिए? हम किसके लिए जी रहे हैं? हमारा उद्देश्य क्या है? सरकार और समाज हमारे साथ कब और कहाँ खड़ा हुआ है? हम अपने लिए कब जियेंगे? कब ये समझेंगे कि दूसरों का मुँह ताकने से कुछ नहीं होगा! हमारी अपनी खुशी का मोल बस हमारे ही हाथ में है! हमारी जिंदगी की सुरक्षा की जिम्मेदारी हमें स्वयं ही लेनी होगी क्योंकि किसी पर दोषारोपण करने से जाने वाला लौटकर नहीं आ सकता!

देश के गौरव का 'भार' उठाती बेटियाँ

डुबकियां सिंधु में गोताखोर लगाता है
जा जाकर खाली हाथ लौटकर आता है
मिलते नहीं सहज ही मोती गहरे पानी में
बढ़ता दुगना उत्साह इसी हैरानी में
मुट्ठी उसकी खाली हर बार नहीं होती
कोशिश करने वालों की हार नहीं होती

आज अगर आप कवि सोहनलाल द्विवेदी द्वारा रचित इस कविता को पढ़ेंगे तो आँखों के सामने बस एक ही चेहरा उभरकर आएगा। मीराबाई चानू का चेहरा। वो चेहरा जो टोक्यो ओलंपिक के पहले ही दिन चाँदी का मेडल जीतकर लाया है। वो चेहरा जिसने तमाम देशवासियों की रगों में बहते खून को नई रवानी दी है। वो चेहरा जिसने आज यह समझाया कि स्त्रियाँ घर-गृहस्थी का ही बोझ नहीं उठाती बल्कि देश को वैश्विक पटल पर चाँदी से रंगना भी बखूबी जानती हैं। यह लड़की जिसके गले में आज चंद्रहार है। उसकी आँखों से छलकते मोतियों ने यह भी सिखाया कि कड़ी मेहनत, लगन एवं आत्मविश्वास से पुरानी कसक को कैसे सुख के आँसुओं में तब्दील किया जाता है।

यह लड़की जब कई किलो वजन उठा रही होती है तो उसके पीछे उसकी पुरानी जीतों को दरकिनार करते हुए 2016 में हारने की एक अफसोसजनक कहानी भी साथ चलती है। जब रियो ओलंपिक में उसके नाम के आगे 'डिड नॉट फिनिश' लिख दिया गया था। यह घटना किसी भी खिलाड़ी का मनोबल तोड़ने के लिए काफी है। उस पर खिलाड़ियों के प्रति हम भारतीयों का व्यवहार हमेशा से ही निर्दयी और संवेदना विहीन रहा है। ये जानते हुए भी कि प्रत्येक खिलाड़ी जीतने के लिए ही खेलता है, अभ्यास में दिन रात एक किये रहता है, हम उनके उत्कृष्ट प्रदर्शन के लिए उन्हें आसमान पर बिठा लेते हैं तो निराशाजनक स्थिति में उनकी छवि की धज्जियाँ उड़ा देने में भी गुरेज नहीं करते! इसी हतोत्साह के चलते कइयों का खेल समाप्त हो जाता है, वे दुख और गहरे अवसाद में ऐसे डूबते हैं कि कभी उबर ही नहीं पाते। लेकिन मीराबाई चानू का खेल तो यहाँ से प्रारंभ होता है।

इतिहास साक्षी है कि मीरा समाज और उसके तानों की चिंता नहीं किया करतीं! वे तो अपने प्रेम को पूजती हैं। पितृसत्तात्मक समाज में सिर उठाकर चलने में यकीन रखती हैं। वही इस मीरा ने भी किया। अपने खेल से प्रेम। फिर अथक परिश्रम, एकाग्रचित्त मन

और कुशल प्रदर्शन के दम पर उन्होंने 2017 में वर्ल्ड वेटलिफ्टिंग चैंपियनशिप में एवं 2018 में ऑस्ट्रेलिया के राष्ट्रमंडल खेलों में गोल्ड जीता। कह सकते हैं कि जुझारूपन और साहस की मिसाल बने हमारे इस 'छोटा पैकेट, बड़े धमाके' ने असफलता को धता बताते हुए दोनों हाथों से सोना बटोरा।

एक बार फिर लकड़ी का बोझ सिर पर उठाए उस लड़की की कहानी, आज टोक्यो ओलंपिक में चाँदी के वक्र से सज गई है। जैसे एक पुराना बोझ उसके मन से हट गया है। जीतने के बाद उसने जो नृत्य किया, उसका वो भांगड़ा दिल से निकली यही बल्ले-बल्ले थी कि आज वो निर्भार हुई। उसे देख यूँ महसूस हुआ कि जैसे वो कटु स्मृति को झटक अपने-आप को यही प्रसिद्ध शेर कह रही हो कि-

गिरते हैं शहसवार ही मैदान-ए-जंग में,

वो तिफ्ल क्या गिरे जो घुटनों के बल चले

ओलंपिक 2021 में रजत पदक से देश का खाता खोलने वाली, साइखोम मीराबाई चानू को हार्दिक बधाई एवं देशवासियों की करोड़ों दुआएं एवं स्नेह भी उन तक पहुँचे।

लेकिन चानू की विजयी मुस्कान को एक खिलाड़ी की जीत भर मान लेना काफी नहीं होगा। यह उस हौसले की भी जीत है जो पराजय के अँधेरे को परास्त करते हुए आपके मस्तक पर जगमगाता है। यह रेगिस्तान हुए बंजर मन में नखलिस्तान की उम्मीद का लबालब भरना भी है। यह संघर्षों की कठिन ऊंची पहाड़ी के उस पार दिखती, मीठी झील की चमचमाती मुस्कान है।

आवश्यक है 'घर संभालने' की अनिवार्य शर्त से मुक्ति

'लड़की घर भी संभाल लेगी, और मैडल भी जीत लेगी'। टोक्यो ओलंपिक में भारतीय महिला एथलीटों के उत्कृष्ट प्रदर्शन से उत्साहित लोग सोशल मीडिया पर कुछ इस तरह की प्रतिक्रिया दे रहे हैं। मनोबल बढ़ाने के उद्देश्य से कही गई ये बात तथ्यात्मक रूप से बिल्कुल सही है। लड़कियों के लिए हमारे समाज में जिस तरह का वातावरण मौजूद है, उन्हें अपने सपनों को पूरा करने के लिए कई सामंजस्य बैठाने पड़ते हैं। 'घर संभालने' की अदृश्य जिम्मेदारी का दबाव उनमें से एक है। फिलहाल इस बात को मैं महिला-पुरुष वाली डिबेट में नहीं ले जाना चाहती। और ओलंपिक में पदक जीतने के साथ देश का मान बढ़ा रही बेटियों के आनंद में अपनी खुशी को महसूस करना चाहती हूँ। क्योंकि, इन खिलाड़ियों के पूरे होते सपने लाखों-करोड़ों लड़कियों को सपने दे रहे हैं।

हम भारतीय इस बात को भलीभाँति जानते और समझते हैं कि इस देश की स्त्रियों में प्रतिभा और लगन की कोई कमी नहीं उनकी क्षमता अपार है और अवसर प्राप्त होते ही वे यह सिद्ध भी कर देती हैं। उनके नाम छपते ये प्रशस्ति-पत्र इसी तथ्य को प्रमाणित करने का उपक्रम भर हैं। एक स्त्री की तो दिनचर्या ही यही है कि वह कई मोर्चों पर एक साथ खड़ी होती है और सफलता प्राप्त करती है। बहुकार्यण एवं समय प्रबंधन में उसका कोई सानी नहीं!

हाँ, कामकाजी हो या घरेलू महिला, घर-गृहस्थी उसकी प्राथमिकता में सदा ही रहे हैं। या ये समझिए कि समाज की सीख ही ऐसी है कि पुरुषों ने ऐसे उत्तरदायित्वों को स्त्रियों के भरोसे रख छोड़ा है। सदियों से हमारी परंपरा भी तो यही रही है कि पुरुष बाहर काम करते हैं और स्त्रियाँ घर संभालती आई हैं। जब समय परिवर्तित हुआ और स्त्रियों ने घोंसले के बाहर निकल अपने पंख पसारना प्रारंभ किये, तब भी शर्त यही थी कि उनकी अपनी उड़ान में घर पीछे नहीं छूटना है।

स्त्रियों को यह भी पता है कि इस उड़ान में सफल होना उनके लिए कितना आवश्यक है वरना जगहँसाई से पहले, घर की स्त्रियों के कटाक्ष ही उनके पर कतरने को पर्याप्त हैं। इसलिए वह जब एक लक्ष्य साधने का दृढ़ निश्चय करती है तो उससे पहले अपने हौसले और सक्षमता को सौ बार तोलती-परखती भी है। सामाजिक परिदृश्य में उसका स्थान कहाँ निश्चित किया गया है, इस तथ्य से भी वह परिचित है ही।

सर्वविदित है कि पितृसत्तात्मक समाज में स्त्री का सिर उठाकर चलना, महत्वपूर्ण मुद्दों पर बोलना और अपनी स्वतंत्रता की बात करना तक कुछ लोगों को खल जाता है। हम उस देश के निवासी हैं जहाँ आज भी कुछ स्थानों पर मात्र अपनी पसंद के कपड़े पहनने से लड़कियाँ मार दी जाती हैं। दूसरी जाति में या मनचाहे लड़के से प्रेम-विवाह परिवारों को हजम नहीं होता! जहाँ बेटी का जन्म ही बोझ लगने लगता है या फिर भ्रूणावस्था में ही उसे 'गिरा' दिया जाता है। जहाँ दहेज के नाम पर उसकी हत्या होती है, रात में निकलने पर उसके साथ बलात्कार किया जाता है। उस पर लड़कों की गंदी निगाह का कारण उनकी विकृत सोच नहीं बल्कि लड़की का पहनावा सिद्ध कर दिया जाता है। वहाँ यदि लड़कियों का प्रदर्शन अच्छा होगा तो प्रशंसा तो होगी ही। क्योंकि हमको पता है कि इनका जीवन एक अंतहीन बाधा-दौड़ से कम नहीं!

जब हम कहते हैं कि 'लड़कियों ने फिर बाजी मारी' तब यह दोनों लिंगों के बीच कोई प्रतिस्पर्धा मानकर नहीं कहा जाता। बल्कि यह तो उन परिस्थितियों को ध्यान में रखते हुए कहा जाता है, जहाँ हमें पता होता है कि आज भी कई लड़कियाँ, घरेलू काम में हाथ बँटाते हुए और तमाम कठिनाइयों से जूझते हुए इतना सब कर रही हैं।

महिला खिलाड़ियों का उत्साह बढ़ाते हुए भी जो पंक्ति लिखी गई है, उसमें भी प्राथमिकता घर संभालने को ही मिली है। क्या लड़कों का उत्साहवर्धन करते हुए भी यही कहा जाता है? नहीं, क्योंकि वे मुक्त हैं हर भार से।

ऐसे समाज से जब कोई स्त्री ओलंपिक में पदक जीतकर लाती है तो उसकी पीठ थपथपा इस प्रजाति पर गर्व करना स्वाभाविक है और आवश्यक भी। जिससे उन सबको प्रोत्साहन मिले, जिनके हृदय में अभी कहीं कोई अचकचाहट है। उन्हें भी एहसास हो कि राह उतनी भी कठिन नहीं, जितना वे माने बैठी हैं। बस पहला कदम रखने भर की देर है। अभी लड़कियों का मनोबल ऊँचा करने का समय इसलिए भी है क्योंकि आत्मविश्वास पाने का संघर्ष वे ही कर रही हैं। जब ये सब आम हो जाएगा। देश के हर राज्य से ऐसी खबरें प्रायः आया करेंगी और हमें इसकी आदत हो जाएगी, तब ही इस तरह के मुख्य शीर्षक बनने पर विराम लगेगा। लेकिन इस प्रक्रिया से पुरुष रूपी राजा को विचलित नहीं होना है। उसकी जगह सुनिश्चित है। बस, रानी के लिए अपने साथ एक सिंहासन और बनवा लेना है।

आपने भी तो कितनी बार ऐसी शीर्ष पंक्तियों को हमेशा ही आश्चर्य मिश्रित हर्ष के साथ देखा-पढ़ा होगा, जहाँ लिखा हो–

'अब लड़कियाँ भी चलाएंगी विमान', 'भारत की पहली महिला ट्रक ड्राइवर'। पेट्रोल पंप में काम करती लड़कियों को देखकर भी यकायक चौंक जाते हैं न! क्योंकि ये सब काम तो हम पुरुषोचित मानकर बैठे हैं! लड़कियों ने भारोत्तोलन और मुक्केबाजी जैसे खेलों में पदक जीते हैं। जो शारीरिक क्षमता भी प्रदर्शित करते हैं। अब कल्पना करो कि प्रायः अपने घरों में जब वजन उठाने का कोई काम होता है तो पिता, पति, भाई या बेटे को बुलाया जाता है। क्या कोई परिवार अपनी लड़की के बचपन में ये सोच सकता है कि ये बड़ी होकर पहलवानी या मुक्केबाजी करेगी! उसे तो हम सौम्यता की मूरत बनाते हैं। लेकिन अब ये सूरत बदल रही है।

इस बदलती सूरत का शानदार नजारा यह है कि टोक्यो ओलंपिक में भारोत्तोलन में मीराबाई चानू ने शानदार प्रदर्शन कर देश को पहला पदक दिला दिया। बैडमिंटन स्पर्धा में पीवी सिंधु दो बार ओलंपिक पदक जीतने वाली पहली भारतीय महिला बनी। एक तरफ भारतीय पुरुष हॉकी टीम सेमीफाइनल में पहुँची, उधर भारतीय महिला हॉकी टीम ने भी ओलंपिक में पहली बार सेमीफाइनल में जगह बनाकर इतिहास रच दिया है। पूल चरण में संघर्षरत यह टीम बाद में पूरी शक्ति के साथ ऐसा खेली कि सब देखते रह गए। इसकी कप्तान रानी रामपाल से लेकर गोलकीपर सविता तक लगभग सभी की ओलंपिक तक पहुँचने की कहानी काफी संघर्ष भरी रही है।

मैरीकॉम ने बॉक्सिंग चुना और देश को इसमें पहले मैडल भी दिला चुकी हैं। इस बार भी उनका प्रदर्शन अद्भुत रहा। मैडल भले ही फिसल गया परंतु काँटे की टक्कर दी है उन्होंने। मैरीकॉम ने 2012 के लंदन ओलंपिक में मुक्केबाजी में कांस्य पदक जीता था। और इस बार लवलीना बोरगोहेन का भी पदक जीतना तय है। इन दोनों मुक्केबाजों की कहानी भी प्रेरणा का विषय है क्योंकि उनके परिवारों ने अपनी बेटियों के खेल को प्राथमिकता दी। मैरीकॉम ने बॉक्सिंग में ऊँचा मुकाम पाने के बाद जीवन साथी चुना। उसने खेल और मैडल को लक्ष्य बनाया यदि उसके दिमाग में घर संभालने का ख्याल होता तो शायद आज वो वहाँ नहीं जा पाती, जहाँ गई हैं।

इस बात से भी मुँह नहीं मोड़ा जा सकता कि अन्य राज्यों की अपेक्षा उत्तर-पूर्वी राज्यों में स्त्रियों की स्थिति बेहतर है। लेकिन दक्षिण और हरियाणा की लड़कियाँ भी कमाल कर रही हैं। कह सकते हैं कि खेलों का जहाँ-जहाँ मान है, वे संस्कृति में ढल गए हैं और लड़कियों अपने निर्णय स्वयं लेने में सक्षम हैं वहाँ से उम्मीद की किरणें प्रस्फुटित हो रही हैं। लड़कियों के मन से घर संभालने का दबाव निकालने का यही समय है। अभिभावकों को भी प्रेरणा लेनी चाहिए कि उनके बच्चों को यदि खेल में रुचि है, तो वे उन्हें

प्रोत्साहित करें। टोक्यो ओलंपिक की उपलब्धियाँ इसी काम में आनी चाहिए।

एक बात और, कभी-कभी गरीब पृष्ठभूमि वाले लड़के-लड़कियों के लिए स्पोर्ट्स नौकरी पाने का जरिया होता है। लेकिन, इक्कीसवीं सदी की लड़कियों के सपने बड़े हैं। वे नौकरी से आगे ओलंपिक मैडल को लक्ष्य बना रही हैं।

'हमें बेटी बचाओ' को एक कोरा आह्वान नहीं समझना है। बेटियों के सपनों को बचाना है, उन्हें उनका अपना आसमान चुनने का अधिकार देना है। 'बेटी पढ़ाओ' की जब बात करें तो चुपके से उनका मन भी पढ़ लेना है। उनकी आँखों में सुख की इक मीठी झील भी छोड़ देनी है। यकीन मानिए जिस दिन इस देश की सारी बेटियाँ अपनी मर्जी से जी सकेंगीं, अच्छे दिन भी आ जाएंगे। हमारी परियों को थोड़ा-सा आसमान तो दीजिये, वो उस पर मैडल टांक देंगी। ओलंपिक तो क्या हम दुनिया की किसी भी पदक तालिका में इतना पीछे नहीं रहेंगे कि इकाई में प्राप्त पदकों पर ही अपने भाग्य को सराहने लगें। लेकिन हमारे जिन खिलाड़ियों ने हमें ये गौरवशाली पल दिए और वहाँ तक पहुँचे, उनको इस देश का सौ-सौ बार सलाम।

पुरुषों के भरोसे न रहकर, सीखनी होगी आत्मरक्षा

राष्ट्रीय अपराध रिकॉर्ड ब्यूरो (एन. सी. आर. बी.) द्वारा बुधवार 15 सितम्बर 2021 को जारी आंकड़े चौंकाने वाले हैं। इन आंकड़ों के अनुसार वर्ष 2020 में महिलाओं के प्रति अपराध की 3,71,503 घटनाएं दर्ज की गईं। ध्यातव्य है कि यह लगभग पौने चार लाख तो मात्र दर्ज हुई संख्या है। वास्तविक घटनाओं का तो अब भी हिसाब नहीं! स्वयं पर हुए अत्याचार की शिकायत करने का हौसला अभी स्त्रियों ने जुटा ही कहाँ पाया है!

बलात्कार के इन शर्मनाक मामलों में राजस्थान सबसे आगे है। 2020 में देश भर से 28,046 रेप केसेस रिपोर्ट किये गए। इनमें से 5,310 राजस्थान से हैं। उसके बाद उत्तर प्रदेश एवं मध्य प्रदेश की गणना ने शर्मसार किया है। बलात्कार के दर्ज प्रयासों की संख्या 3741 है। यहाँ भी 965 मामलों के साथ राजस्थान शीर्ष पर दिख रहा। उसके बाद पश्चिम बंगाल एवं असम का नाम है। महानगरों की श्रेणी में दिल्ली के बाद जयपुर में सर्वाधिक बलात्कार हुए। जयपुर के बाद मुंबई है। महिलाओं के विरुद्ध अपराध के मामलों में उत्तर प्रदेश सबसे आगे है। तत्पश्चात पश्चिम बंगाल एवं राजस्थान आते हैं। चेहरे और प्रदेश के नाम भले बदलते रहेंगे पर अपराध की वीभत्सता हर जगह अपने चरम पर है।

हम अफगानी स्त्रियों की दुर्दशा पर तो भीषण दुख मना रहे हैं लेकिन अपने देश में स्त्रियों के प्रति हुए अपराध को देख घड़ियाली आँसू तक बहाने का समय नहीं निकाल पाते! क्या समाचार-पत्र में छपे ये आँकड़े हम सबके मुँह पर करारे तमाचे की तरह महसूस नहीं होते? क्या हम इन अपराधों के अभ्यस्त हो चुके हैं? क्या हमें इनसे सचमुच कोई फ़र्क़ नहीं पड़ता? क्योंकि यदि हमें कोई फ़र्क़ पड़ता तो हमने अब तक सत्ता की ईंट से ईंट बजा दी होती! हम चुप हैं क्योंकि हमने बलात्कारियों को जमानत पे छूटते देखा है। छूटने के बाद पीड़िता के परिवार को तबाह करते भी देखा है। हमने यह मान लिया है कि हमारा तंत्र यानी 'सिस्टम' ही ऐसा है जहाँ सरेराह हुए अपराध को प्रमाणित करते-करते पीड़ित पक्ष स्वर्ग सिधार जाए पर न्याय नहीं मिलता! आखिर नक्कारखाने में तूती की आवाज कौन सुनेगा!

उस पर बलात्कार के मामलों में हमारे नेताओं के वचन इस अपराध को और हल्का करते आए हैं। 'लड़कों से तो गलती हो ही जाती है', 'रात को बाहर निकलने की क्या जरूरत थी?', छोटे-छोटे कपड़े पहन निकलेंगी, तो पुरुष उत्तेजित होगा ही'। तात्पर्य यह कि स्त्रियों को उपभोग की वस्तु समझने वाले कुत्सित मानसिकता से घिरे ये लोग यह मानकर ही चल रहे हैं कि पुरुष तो सदा से ही ऐसा

रहा है अतः स्त्रियों के रेप का कारण स्वयं स्त्रियाँ ही हैं! ये चाहते हैं कि स्त्रियाँ घर से न निकलें, निकलें तो इनके हिसाब के वस्त्र पहनकर निकलें और सूर्यास्त से पहले घर आ जाएं। यदि इनसे यह प्रश्न किया जाए कि फिर छोटी मासूम बच्चियों, तथाकथित 'सभ्य' वस्त्रधारी लड़कियों के साथ, दिन के समय हुए बलात्कार पर आपका क्या कहना है तो ये अश्लीलता से हँसते हुए बगलें झाँकने लगते हैं। इनकी तालिबानी सोच पर धिक्कार तो है ही, साथ ही एक राय भी कि इतना सब करने से अच्छा है कि आप पुरुष लोग घर बैठें क्योंकि हम स्त्रियाँ इस दुनिया को बेहद खूबसूरती से संभाल सकतीं हैं।

लेकिन इन भावनात्मक बातों से ऊपर भी एक मुद्दा है, महिला सुरक्षा का! महिला ही क्यों बल्कि 'मानव-सुरक्षा' का। हमने इसके लिए आज तक क्या किया है? विद्यालयों में कला, वाणिज्य, विज्ञान से जुड़े सारे विषय पढ़ाए गए। कुछ दशक पहले तक खेल एवं नैतिक शिक्षा विषय भी हुआ करते थे। लेकिन आत्म-रक्षा कैसे की जाती है, यह हमें कभी नहीं सिखाया गया। घर में भी यही सिखाया जाता रहा कि कोई हम पर हमला करे तो वो जैसा बोले करते जाओ। अपना सारा सामान उसे दे दो क्योंकि जान है तो जहान है! बलात्कार के समय आरोपी के सामने गिड़गिड़ाना, उसके पैर पड़ना, जीवन बर्बाद न करने के लिए रो-रोकर भीख मांगना ही महिलाओं की नियति बनी रही। उसके बाद भी दर्द को समझने कोई नहीं आया और उस महिला को तिरस्कृत कर उस पर ही दोष मढ़े गए। कुछ स्थानों पर जहाँ इज्जत का मापदंड योनि सुरक्षा है, वहाँ तो उसे आत्महत्या के लिए उकसाने वाले मामले भी देखे गए हैं।

हमारे समाज में हर स्थान पर लड़कियों को अकेले भेजने की बजाय उनके साथ कोई लड़का भी भेज दिया जाता है। भले ही वो पिद्दी सा हो! यह भी कोई तरीका हुआ! उस पर कामकाजी स्त्रियों के लिए यह संभव भी नहीं कि हर समय एक भाई या मित्र के साथ ही रहें। क्या अब समय नहीं आ गया है कि हम अपने अधिकार और रक्षा के लिए किसी पुरुष की प्रतीक्षा न कर स्वयं दुश्मन पर वार करना सीखें! अपनी रक्षा स्वयं करें! लेकिन ये जोश में कहने भर से नहीं होगा, करना होगा! हमें इसका प्रशिक्षण लेना होगा। हमारे बच्चों को इसको सीखने की शुरुआत विद्यालय स्तर से करनी होगी! हमें हमारे बच्चों के लिए ऐसी कोई परियोजना चाहिए, जहाँ प्रशिक्षित होकर ये पूरे आत्मविश्वास के साथ आने वाले हर खतरे से जूझने में सक्षम बन सकें। अपराध का मुँहतोड़ जवाब देने के लिए हमें शरीर के साथ-साथ अपनी मानसिकता भी मजबूत करनी होगी।

'सिस्टम' से उम्मीद करना व्यर्थ है। स्त्रियों के प्रति बढ़ते अपराध के आंकड़ों को, आत्मरक्षा के गुर सीखकर हम स्त्रियाँ ही रोक सकते हैं।

एक चिट्ठी, इस दुनिया की हर आयशा के नाम

मेरी प्यारी आयशा

अभी जबकि लगभग एक सप्ताह बाद पूरा देश मिलकर, महिलाओं को उनके दिवस की बधाई देने में जुट जाएगा, इस बीच तुमने साबरमती में कूदकर अपनी जान दे दी है। इतना ही नहीं बल्कि आत्महत्या से पहले एक वीडियो भी बनाया है जिसका एक-एक शब्द झकझोर कर रख देता है। जीवन के आखिरी पलों के दौरान भी अभिवादन की तुम्हारी तहजीब देखकर मैं हैरान हूँ। तुम अंत में थैंक यू कहना भी नहीं भूलतीं। लेकिन सच कहूँ तो तुम्हारी निडर आवाज और मुस्कान ने डरा दिया है मुझे। न जाने वो कौन से पल रहे होंगे कि तुमने मृत्यु को सुख पाने का मार्ग समझ लिया। मुझे बेहद दुःख और अफसोस है कि हम इस दुनिया को तुम्हारे जीने लायक बनाने में विफल रहे।

हम एक ही शहर के हैं पर कभी मिले नहीं मेरा तुमसे जो नाता है वो साबरमती का है, रिवर फ्रंट का है। ये जगह मुझे हमेशा से सुकून भरी लगती रही है। तुमने भी अपने वीडियो में सुकून की बात कही है। पर कितना फ़र्क़ है तुम्हारे और मेरे सुकून में! तुम कहती हो 'मैं हवाओं की तरह हूँ, बस बहना चाहती हूँ और बहते रहना चाहती हूँ। किसी के लिए नहीं रुकना'। और मैं चाहती हूँ कि काश तुम नदी के पानी से अठखेलियाँ करतीं और जीवन धार में बहती जातीं। हर संघर्ष का सामना करतीं। ये जीवन तुम्हारा अपना है, तुम्हें इसे पूरा और बेबाकी के साथ जीना था।

तुम्हारे चेहरे पर सहज मुस्कान है लेकिन आँखों में दर्द का इक सैलाब भरा है। एक ऐसा सैलाब, जिससे हर स्त्री किसी-न-किसी रूप में जरुर रिलेट कर सकेगी। तुम्हारे जो शब्द हैं, वो हमारे मस्तिष्क पर तमाचे की तरह पड़ते हैं। तमाचा, जो बार-बार यही याद दिलाता है कि तुम में और आयशा में कोई फ़र्क़ नहीं! लेकिन ये बताओ कि इतनी मानसिक पीड़ा के बावजूद भी तुमने महान बनने का वह नैसर्गिक गुण क्यों नहीं छोड़ा? जिसे हम स्त्रियों ने सदियों से किसी मंगलसूत्र की तरह दिल से लगाकर रखा है। तुम्हें एक बार को भी नहीं लगा? कि जब तक स्त्रियाँ अन्याय को सहती रहेंगी, उसके विरोध में आवाज उठाने की बजाय चुप्पी साध लेंगी और उफ तक न करेंगी, तब तक उनकी असमय मृत्यु का ये दौर अनवरत जारी रहेगा! तुम्हें बहुत हिम्मत दिखानी थी, गुड़िया।

आयशा, तुम्हारी बातें सुन जितनी पीड़ा हुई है, उतना ही क्रोध भी आया है। यूँ भी दिल किया कि तुम्हें डांटकर कहूँ 'पागल लड़की! तुम्हें निकल आना था, उस जहन्नुम से! डूबते हुए भी तुम

एक दहेज लोभी इंसान को बचाना चाहती हो? उसे बरी करना चाहती हो? ये कैसी मोहब्बत है तुम्हारी कि जिसने तुम्हारा जीवन बरबाद कर दिया, तुम अपनी जान देकर उसे बचा रही हो?' तुम्हारे पापा-मम्मी ने तुम्हें कितना रोका, कसमें दीं, मिन्नतें कीं। यहाँ तक कि दहेज का केस वापिस लेने को भी तैयार हो गए, तब भी तुम हार गईं? तुम्हें नहीं पता कि तुम कितनी भाग्यशाली थीं कि तुम्हारे माता-पिता तुम्हारे साथ थे।

पता है, तुम केवल मोहब्बत हो! और तुम्हें खोने वाले अभागे। तुम अपने पिता से आग्रह करती हो कि 'कब तक लड़ेंगे अपनों से? आयशा लड़ाइयों के लिए नहीं बनी। प्यार करते हैं उससे, उसे परेशान थोड़े न करेंगे। अगर उसे आजादी चाहिए, ठीक है वो आजाद रहे'।

काश! तुम्हारी ये बात दुनिया समझ ले तो हर चीज सुंदर हो जाए। जो तुम ठहरतीं तो दुनिया आसानी से समझ पाती।

जानती हो, तुमसे कोई गलती नहीं हुई थी और न ही तुम में या तुम्हारी तकदीर में कोई कमी थी। बस, तुम अपने-आप को नहीं जान पाईं। तुम उन अपनों को भी नहीं देख पाई जो तुमसे बेपनाह प्यार करते हैं। तुमने अपना जीवन एक ऐसे इन्सान की खातिर गँवा दिया, जिसे पैसे से प्यार था। लेकिन सारे इन्सान बुरे नहीं होते!

हाँ, तुम्हारी ये बात सच है कि 'मोहब्बत करनी है तो दोतरफा करो एकतरफा में कुछ हासिल नहीं!' इसमें एक बात और जोड़ती हूँ कि मोहब्बत दोबारा भी हो सकती है। ये बात याद रखना अब।

तुम्हारा आखिरी फोन कॉल दिल दहला देने वाला है, आयशा। तुम्हारे आँसुओं ने बेहद रुलाया है। तुम्हारे माता-पिता का सोचकर भी कलेजा कांप उठता है। मैं उस पिता की निरीह अवस्था और हताशा को सोचती हूँ जिसे फोन पर पता चलता है कि अगले ही पल उसकी बेटी नदी में छलांग लगाने वाली है। उस माँ के दर्द को समझने की नाकाम कोशिश करती हूँ जो बार-बार तुमसे रुकने का अनुरोध कर रही है। मासूम लड़की, तुम्हें परिस्थितियों का डटकर सामना करना था। अपने आप को किसी से कम नहीं आंकना था। तुम अपार संभावनाओं से भरा चेहरा थीं।

तुम्हारी विदाई का गुनहगार ये समाज भी है जो तुम्हारे लिए ऐसा वातावरण ही नहीं बना पाया कि तुम स्वयं को अकेला न महसूस कर सको। या फिर इसे तुम्हें अकेले चलना सिखा देना चाहिए था। तुम्हें बता देना चाहिए था कि तुम्हारी मुस्कान, तुम्हारे जीवन की तरह कितनी अनमोल है।

मैं तुम जैसी तमाम लड़कियों से कहना चाहती हूँ कि आखिर

हम क्यों अपने सपनों और खुशियों को किसी और के जीवन से जोड़ें? हम क्यों न अपने हक की लड़ाई खुद लड़ें? किसी से इतनी अपेक्षा क्यों रखें कि उनके पूरा न होने पर हम ही टूट जाएँ। आत्महत्या, तो अवसाद की पराकाष्ठा है। हम इस राह से बाहर निकल, आत्मसम्मान के साथ जीना क्यों न चुनें? सब अच्छे लोग यूँ दुनिया को छोड़ देना चुन लेंगे, तो इसे संवारेगा कौन?

तुम जहाँ भी हो, हर खुशी तुम्हारे साथ हो। स्नेह तुम्हें।

सुसाइड करने वाली हर तीसरी महिला, भारतीय ही क्यों?

आयशा की आत्महत्या ने फिर से वही सारे प्रश्न खड़े कर दिए हैं जिनसे हम बार-बार बचकर निकलना चाहते हैं। लेकिन क्या इस तरह की घटनाओं और आत्महत्या का जिम्मेदार वो समाज नहीं? जो बचपन से ही एक लड़की के दिमाग में यह कूट-कूटकर भर देता है कि 'वह फूलों की डोली में बैठकर जिस घर में प्रवेश करेगी, उसकी अर्थी भी वहीं से उठेगी!' क्यों उठे उसकी अर्थी वहीं से? यदि वह उस घर में खुश नहीं तो क्या वो उस घर को छोड़कर नया जीवन नहीं जी सकती? कब तक वो अभिशप्त जीवन जीती रहे?

उसे यह भी समझाया जाता है कि 'मारे या पीटे पर पति तो परमेश्वर होता है। थोड़ा तो सहन कर ही लेना चाहिए'। हाँ, बिलकुल कर लेगी लेकिन तब ही, जब पति भी उसके हाथ का थप्पड़ खाकर बर्दाश्त करना सीख जाए और चुप रहे। पढ़ते ही कैसे धक् से चुभ गई न ये बात?

यही होता है कि जब तक स्त्री सहन कर रही है, सब अच्छा है। वो मर भी गई, तब भी हम उन सारे कारणों को ढूँढने में लग जाते हैं जो हमें दोषमुक्त कर दे। लेकिन कब तक? आयशा तो एक नाम भर है जो हमारे समाज की स्त्रियों का प्रतिनिधित्व करता है। लेकिन आँकड़े जो सच्चाई बयान कर रहे हैं, उनसे आखिर कैसे मुँह फेरा जा सकता है?

भारत में होने वाले तलाक की दर दुनिया में सबसे कम है–

प्रतिवर्ष यूनाइटेड नेशंस, ग्लोबल डिवोर्स रेट दर्ज करता है। इसके अनुसार पूरी दुनिया में भारत में होने वाले तलाक की दर सबसे कम है। यहाँ 1000 शादियों में से केवल 13 में तलाक होता है। यह आँकड़ा डेढ़ प्रतिशत से भी कम है। जबकि स्पेन, फ्रांस, यूनाइटेड स्टेट में सर्वाधिक मामले दर्ज किये गए हैं।

लेकिन इसका मतलब ये कतई नहीं कि हमारे यहाँ सभी खुशहाल वैवाहिक जीवन जी रहे हैं। दरअसल 'हैप्पी मैरिज लाइफ' एक मिथ्या है। हम लोग तोड़ने से बेहतर, बर्दाश्त करने को मानते हैं। समझौता मंजूर है हमें, पर अलग होना नहीं!

जो स्त्रियाँ वैवाहिक जीवन से तंग आकर आत्महत्या करती हैं, उन्हें पता है कि इस देश में तलाक लेकर जीना आसान नहीं अकेली स्त्री को उपभोग की वस्तु की तरह देखा जाता है। वो अपने पिता और पति के अलावा और किसी पर आसानी से विश्वास नहीं कर पाती। गृहिणियों को आर्थिक असुरक्षा का भय भी सालता है। ऐसे में उन्हें यही सही लगता है कि जैसे-तैसे इसी चारदीवारी में

जीवन काट दें। वे तलाक के स्थान पर, सहजता से घुटन भरे जीवन में रहने के विकल्प को चुन लेती हैं। वे पितृसत्तात्मक समाज के इस सूत्र को अपना ध्येय वाक्य बना बैठती हैं कि 'सुहागन मरना सौभाग्य की निशानी है'।

दुनिया में सुसाइड करने वाली हर तीसरी महिला भारतीय है–

विश्व की महिला जनसंख्या में भारतीय महिलाओं की हिस्सेदारी 18 प्रतिशत है, जबकि महिलाओं की आत्महत्या के मामलों में यह 36 प्रतिशत तक बढ़ जाती है। यानी घुट-घुटकर मरने वाली स्त्री, एक दिन मौत को ही चुन लेती है। दरअसल भारतीय समाज की मानसिकता ऐसी है कि विवाह के बाद बेटी पराया धन मान ली जाती है। मायके में उसका स्वागत तो सदैव होता है लेकिन यदि वो किसी परेशानी के चलते पति का घर छोड़कर आना चाहे तो उसके अपने ही, उसके लिए दरवाजा बंद करते नहीं हिचकते। उसे समझौते के लिए समझाया जाता है कि थोड़ा बर्दाश्त कर, सब ठीक हो जाएगा। समाज के सामने अपनी नाक का हवाला दिया जाता है कि तू मायके आकर बैठ गई तो लोग सवाल करेंगे। लड़की इशारे समझ जाती है और फिर कभी अपनी पीड़ा नहीं बाँटती।

हर परिवार ऐसी ही संस्कारी लड़की और बहू की अपेक्षा रखता है। 'लड़का चाहे कुछ करे, उसका चल जाएगा। पर तू तो लड़की है!'

'मैं तो लड़की हूँ!' यही बात उस लड़की के दिल में भी गहरे बैठ जाती है और फिर वो अपनी महानता, त्याग और बलिदान की पुस्तक भरने के लिए खुद ही अपना सिर, कलम करवाने को तैयार रहती है। क्योंकि वो तो लड़की है! उसे तो महान बनना ही होगा! दुनिया के सामने महान बनते बनते ये लड़की अपनी ही नजरों में रोज गिरती है। सबको खुश रखने वाली ये लड़की, अकेले में रोज रोती है।

हमारे तथाकथित संस्कार हमें हर दर्द को सहने की शक्ति सिखाते ही आये हैं। ऐसे में स्त्रियों के पास उसी नर्क में रहने के अलावा और कोई चारा नहीं होता! और फिर एक दिन जब पानी सिर से ऊपर निकलने लगता है तो वे हारकर इस दुनिया को अलविदा कह देती हैं।

सबसे अधिक सुसाइड के मामले 15 से 29 आयु वर्ग वाली युवतियों के–

नई पीढ़ी में आत्महत्या की बढ़ती दर बेहद चौंका देने वाली है। प्रेम, बेमेल विवाह, उनके प्रति होने वाले अपराध के चलते युवतियां भावनात्मक रूप से बेहद कमजोर एवं असहाय महसूस करती हैं। कई बार नौकरी में होने वाली परेशानियों से भी वे तंग

आ जाती हैं। भविष्य के लिए देखे गए सुन्दर सपनों की तस्वीर, जब उनके वर्तमान से मैच खाती नहीं दिखती तो वे हताश हो जाती हैं। पापा की परी के भीतर जैसे ही सामाजिक चेतना जागृत होती है, उसके सामने कई भयावह चुनौतियां आ खड़ी होती हैं। उसमें लड़ने की हिम्मत और जज़्बा तो होता है पर धैर्य नहीं कई बार इनके लिए विद्रोह का तरीका मर जाना ही होता है। वे इतने गहरे अवसाद में चले जाते हैं कि फिर लौट ही नहीं पाते। दोष सामाजिक और पारिवारिक वातावरण का ही है क्योंकि हमने अपने बच्चों को जीतना सिखाया है, सफलता पाने की सारी किताबें रटा रखी हैं लेकिन असफलता से जूझने का पाठ कभी नहीं पढ़ाया। उनके सिर पर हाथ फेरकर कभी नहीं कहा कि हारना भी उतनी ही स्वाभाविक प्रक्रिया है जितनी कि जीतना।

बढ़ती आत्महत्या के कई कारण हैं–

बड़ी-बड़ी डिग्रियाँ हासिल कर लेने के बाद भी स्त्रियों में सशक्तिकरण का अभाव है। वे अपनी हर खुशी को किसी एक व्यक्ति की उपस्थिति या अनुपस्थिति से जोड़कर देखती हैं। अपना सारा जीवन उसी के इर्दगिर्द समर्पित कर देती हैं और जब एक दिन उससे अलग होने की सोच भी सामने आती है तो वे इस स्थिति के लिए मानसिक रूप से तैयार नहीं हो पातीं। उन्हें पुरुष के बिना रहना आया ही नहीं

पितृसत्तात्मक समाज में घरेलू स्तर पर वे, प्रारंभ से ही निचले दर्जे पर खड़ी नजर आती हैं। अपने जीवन से जुड़े हर निर्णय के लिए बचपन में पिता, भाई, फिर पति और अंत में बेटे पर निर्भर। एक परजीवी सा जीवन उन्हें तथाकथित संस्कारों का संरक्षण करना ही सिखाता आ रहा है। अत्याचार सहना पर उफ न करना, बचपन में भाई और बड़े होकर पति, बेटे को बचाना, उनकी उम्र के लिए व्रत करना। उन्हें लगता है कि पति या भाई के अलावा दुनिया उन्हें नोच खाएगी।

पहले महिलाएं अपने साथ होने वाले दुर्व्यवहार को अपना भाग्य मानकर जीवन काट देती थीं। 'भला है बुरा है, जैसा भी है' 'मेरा पति मेरा देवता है' को ब्रह्मवाक्य बनाकर जी लेती थीं। अब वे शिक्षित हैं और अपने अधिकारों के प्रति सजग भी। जब हल नहीं निकलता तो वे कुंठा से भर जाती हैं। कुछ रास्ता नहीं सूझता तो आत्महत्या कर जीवन समाप्त कर लेती हैं।

कैसा दुर्भाग्य है कि हमने अपनी दुनिया को इतना बदसूरत और असुरक्षित बना डाला है कि स्त्रियाँ यहाँ अकेले जीना ही नहीं चाहतीं। हमें अपनी बेटियों, अपने आसपास की स्त्रियों को सबसे पहला पाठ यही देना है कि वे हर हाल में किसी से कमतर नहीं

हैं। हमें उन्हें मुश्किलों से लड़ना सिखाना है और धोखे से संभलना भी। उनके दर्द को बांटना, उनकी परेशानियों को सुनना भी सीखना है। उन्हें यह विश्वास भी दिलाना है कि हम उनके साथ खड़े हैं।

और लड़कियों, तुम आत्मनिर्भर, निडर और सशक्त बनो। हर बात के लिए किसी का मुँह न ताको। जब तक तुम हो, ये दुनिया सुंदर है। तुम्हारी हार, इस दुनिया की हार है।

जब शादी नर्क, तब स्वर्ग है तलाक

तलाक सहज नहीं होता। उतना सहज तो बिल्कुल नहीं, जितना आमिर खान और किरण राव के साझा बयान से दिखा। तलाक को लेकर ये 'साझापन' कड़वाहट के कई पलों के बाद हासिल होता है। सेलिब्रिटीज की शादी हमें परीकथा जैसी लगती है, और उनके तलाक भी उतने ही करिश्माई। लेकिन तय जानिए कि कोई तलाक खेल के लिए नहीं होता। कि चलो, शादी-शादी खेल के बोर हो गए तो अब तलाक-तलाक खेलते हैं। कड़वाहट भरी जिंदगी अपनी उम्र पूरी करने के बाद ही तलाक के मोक्ष तक पहुंचती है।

कड़वे रिश्ते की पहचान क्या होती है? वहाँ आपको दिखाई देता है बिखराव, शिकायतें, कहासुनी, कभी-कभी सिर फुटव्वल भी। विश्लेषण-प्रेमी समाज इसे अपने ही ढंग से पेश करता है। कभी वो इस कड़वाहट के मजे लेता दिखता है। कभी इसे नियति मानकर निभा ले जाने की सलाह देता है। कभी इसे दो बर्तन के टकराने जैसा सामान्य बना देता है। तो कभी पति-पत्नी में से किसी एक को दोषी ठहराकर अपना पल्ला झाड़ लेता है। वो सब कुछ करता है पर तलाक की इजाजत तो हरगिज नहीं देता। ऐसे समाज का पक्ष भी समझा ही जा सकता है। आखिर उसी ने तो शादी को सात जन्मों के बंधन की संज्ञा दी थी। ऐसे में तलाक तो कुफ्र ही हुआ न!

विवाह यदि जन्म-जन्मांतर की संस्था है तो इसके भी कुछ तय नियम हैं। वे नियम जिनका उच्चारण, सात फेरों के समय चुहलबाजी करते हुए किया जाता है यदि उनका गंभीरता से पालन हो तो इससे खूबसूरत और पवित्र कोई रिश्ता ही नहीं! लेकिन वो सुख-दुख के साथी, हर बात साझा करने वाले दोस्त, परस्पर परवाह के हसीन ख्याल सब उस अग्नि के साथ ही स्वाहा हो जाते हैं या कुछ एक साल निभाकर भुला दिए जाते हैं। एक ही घर में रहने वाले दो इंसान कब एक-दूसरे के लिए अजनबी बन जाते हैं, पता भी नहीं चलता! पत्नी बच्चों और ससुराल में रम जाती है

और पति पैसे कमाने की मशीन बनकर रह जाता है। एक-दूसरे को सुनने-समझने का वक़्त ही नहीं रहता।

अपने भारतीय समाज में विवाह को लेकर एक बड़ी प्रचलित कहावत है कि 'शादी वो लड्डू है, जो खाए सो पछताए और जो न खाए वो भी पछताए!' अब लड्डू है तो उसकी रेसिपी भी अवश्य होगी। जी, बिल्कुल है। यदि इसे 'अटूट बंधन' बनाना है तो फिर उम्र भर एक दूसरे के साथ पूरी ईमानदारी से निभाना चाहिए। उसके लिए परस्पर स्नेह और सम्मान की मिठास बहुत जरूरी है। साथ ही समझदारी का मेवा भी धर दिया जाए तो क्या बात!

लेकिन क्या सचमुच ऐसा होता दिखता है आपको? ताली दोनों हाथ से बजती है लेकिन होता यूँ है कि एक पक्ष निभाते-निभाते थक जाता है। शिकवे-शिकायतों का एक लंबा दौर चलता है। उसके बाद फिर एक वो समय भी आता है जब इस स्थिति को स्वीकार कर दोनों अपनी-अपनी दुनिया में रम जाते हैं। हाँ, गाहे-बगाहे 'समझौता गमों से कर लो' की धुन दिमाग में बज उठती है जिसे व्यस्तता की गहरी खाई में गिराकर रोक दिया जाता है। प्रयास रहता है कि यह ध्वनि और हमारे भीतर का हाल दुनिया तक न पहुँचे।

भले ही हर सुबह कोई खुशी लेकर न आती हो, हर सूर्योदय के साथ मन निराशा के गहरे भँवर में डूबने लगता हो पर जीवन फिर भी गतिमान रहता है। कभी बच्चों के लिए, कभी परिवार के लिए, तो कभी समाज के भय से समझौते कर लिए जाते हैं। क्यों न हों! बाबुल के आँगन से पाँव बाहर धरते समय 'भला है, बुरा है, जैसा भी है मेरा पति मेरा देवता है' का मंत्र लड़की के कानों में पहले से ही फूँक जो दिया गया होता है।

'लोग क्या कहेंगे!' के सूत्र को अपनाते हुए, घर में दिन-रात लड़ने-झगड़ने और एक-दूसरे का जीवन नर्क बनाकर रखने वाला जोड़ा भी सार्वजनिक स्थानों पर 'परफेक्ट कपल' की वज्रदंती मुस्कान लिए फिरता है।

वाकई कितना खोखलापन लिए घूमते हैं हम! दुनिया को खुश दिखाने के लिए जी रहे हों जैसे!

विवाह, हमारी सामाजिक परंपरा का अभिन्न अंग है। जीवनसाथी के सपने इसी से जुड़े होते हैं। भविष्य की खूबसूरत इमारत इसी की नींव पर खड़ी होती है। हमारे घरों में बच्चे के जन्म के साथ ही उसके विवाह की कल्पनाएं सजने-सँवरने लगती हैं। बेटी-बेटे की शादी में ये करेंगे, 'वो करेंगे' के तमाम किस्से आपने भी खूब कहे-सुने होंगे।

तमाम रस्मों को निभाते हुए दो इंसानों के घर की चौखट एक हो जाती है। नए रिश्ते, नए परिवार बनते हैं। दुनिया पहले से कुछ और बड़ी हो जाती है तो साथ में उत्तरदायित्व भी बढ़ जाते हैं। संयुक्त परिवार हो तो दोनों पक्षों की अपेक्षाएं भी बहुत जुड़ जाती हैं।

अब हमारी पूर्व की जो सामाजिक व्यवस्था थी, उसमें दोनों की भूमिका सुनिश्चित थी। अलिखित रूप से यह तय ही था कि पत्नी घर संभालेगी और पति बाहर का काम। घर-गृहस्थी चलाने और पैसे कमाने की दोनों की अपनी-अपनी जिम्मेदारी थी। उसे निभाया भी जा रहा था। उस समय की कामकाजी स्त्रियाँ भी गृहस्थी की जिम्मेदारी से मुक्त न थीं और दोहरी भूमिका निभाती थीं।

लेकिन तब भी ऐसा नहीं था कि हर जोड़ा अपने जीवन से

संतुष्ट ही था। लेकिन असंतुष्टि की स्थिति में भी पुरुष सामाजिक मर्यादाओं और इज़्ज़त के चलते तथा स्त्री आर्थिक एवं सुरक्षा संबंधी पहलुओं के कारण एक-दूसरे का दामन थामे रहे। अब भी इस स्थिति में कोई विशेष परिवर्तन नहीं आया है।

प्रत्येक रिश्ता उतार-चढ़ाव से गुजरता है। असल में हम सबको, खासतौर से स्त्रियों को संस्कारों की घुट्टी इस कदर पिलाई गई है कि चाहे लाख दुख सह लो, पर पति का घर न छोड़ना कभी! 'मायके से डोली उठती है और पति के घर से अर्थी!', 'पति परमेश्वर' जैसे ब्रह्मवाक्य उसके दिमाग में भर दिए गए। भरने की भी जरूरत कहाँ! अपने-अपने घर, परिवार या आस-पड़ोस में सब हमने-आपने देखा ही हुआ है कि भीतर ही भीतर घुटते रहेंगे लेकिन अलग न होंगे!

आपने ऐसे कई वैवाहिक जोड़े देखे होंगे, जहाँ दोनों की अपनी एक अलग-अलग दुनिया है। उन्हें आपस में कोई सरोकार ही नहीं! दुनिया भर से हँसी-ठट्ठा करने वाले ये लोग अपने घर में ठीक से मुस्कुराते भी नहीं

स्त्रियाँ अपना दुख पति से इसलिए नहीं बाँटतीं कि पति को कोई तनाव न हो, घर का वातावरण न बिगड़े, बच्चों पर नकारात्मक प्रभाव न पड़े। पति ये मानकर चलते हैं कि उनकी कार्यालयीन समस्या को समझना पत्नी के बूते की बात नहीं! सो वो भी चुप कर जाता है। दोनों को आपस में कुछ बात बुरी लगे तो ये सोच चुप रह जाते हैं कि बात को क्यों बढ़ाना! ऐसे ही करते-करते तमाम छोटी बातें कब जीवन को कड़वाहट से भर देती हैं, इसका हमें अनुमान ही नहीं होता! रिश्ते में मधुरता बनाए रखने के लिए जरूरी है कि मन में जो भी है, वो कह दिया जाए। यदि कहीं कुछ खटक रहा है तो बता दिया जाए, उसके गांठ बनने की प्रतीक्षा न की जाए। जब बहुत कुछ जमा हो जाता है तो लावा बन निकलता है। वहाँ एक-दूसरे को समझने-समझाने की गुंजाइश खत्म हो चुकी होती है। सामने सब अच्छा लेकिन बॉन्डिंग नहीं हो तो अच्छे से अच्छे रिश्ते भी कई सालों बाद ढह जाते हैं।

बॉन्डिंग ऐसे ही तो होती है। उसके लिए अच्छे के साथ-साथ बुरे पल भी जुड़ जाएँ तो कोई बुराई नहीं बल्कि विषम परिस्थितियों में तो एक-दूसरे के व्यक्तित्व को समझने का सही अवसर मिलता है। अच्छे समय में तो सब अच्छा ही व्यवहार करते हैं। बात तब है जब प्रतिकूल समय में भी साथ खड़े होने और उससे लड़ने का हौसला हो।

उच्च वर्ग और निम्न वर्ग में तो अलगाव की खबरें फिर भी सहजता से निगल ली जाती हैं लेकिन मध्यम वर्ग इस मामले में

भी पिसता आया है। अपनी खुशी का सोचने से पहले वह सौ बार अपनी नाक टटोलेगा, अपने आसपास के लोगों का मुँह देखेगा फिर ठंडी आहें भर स्वयं को देवी-भगवान मान खुश होने का नाटक शुरू कर देगा। खुशियों को मापने के भौतिक पैमाने भी बना ही लिए गए हैं। 'क्या कमी है तुम्हारे पास? सब तो है और क्या चाहिए' कहकर समझाने वाले लोग मन पढ़ना नहीं जानते। वे ये कभी नहीं समझ सकते कि जो एक जीवन हमें मिला है, उसे खुशी से जीने का हक सबको है। दूसरों की खुशी के लिए कोई कब तक कुर्बानी देता रहेगा?

हमारी दिक्कत ये है कि हम तलाक को किसी रिश्ते के टूटने की तरह ही देखते रहे हैं, जिस दिन हम इसे दो इंसानों के जीवन से जुड़ी खुशी की तरह देखने लगेंगे, हमारा चौंकना और उनकी लानत-मलानत करना स्वतः ही बंद हो जाएगा।

आज भी क्यों जरूरी हैं, कृष्ण?

हम आस्थाओं से तरबतर देश के नागरिक हैं। संस्कृति के रक्षक कहलाते हैं और इसे बचाने के लिए कट्टर कहलाये जाने से भी तनिक गुरेज नहीं करते! सनातन परम्पराओं की रक्षा हेतु हम अपने प्राण तजने को भी आतुर हैं (ऐसा बहुत से लोग कहते हैं)।

प्रश्न यह है कि जिस ईश्वर की हम शपथ लेते हैं, जिसे दिन-रात पूजते हैं, जिसकी छवि अपने हृदय में बसाए रखने का भरम पालते हैं फिर उस ईश्वर के आदर्शों को कैसे भूल जाते हैं?

मैं बहुत धार्मिक तो नहीं हूँ पर किसी अदृश्य, अनाम शक्ति के प्रति आस्था अवश्य रखती हूँ। प्रकृति की अपार सुंदरता को जब-जब निहारती हूँ तो सदैव ही उस चेहरे को खोजती हूँ जिसने जल को इतनी निर्मलता दी। वृक्षों को हरियाली देकर उनकी गोदी फल-फूलों से लबालब कर दी, साथ ही उन्हें छाया का हुनर देकर अपनी पूँजी को बाँटना भी सिखा दिया। नदियों को तमाम झंझावात के मध्य भी कलकल करते हुए बहना सिखाया। पर्वतों को हमारा प्रहरी बनाया और सूर्य में जीवन तत्व भर दिया। उसके बाद इन सभी की आपस में ऐसी मित्रता कराई कि सृष्टि के निर्माण से लेकर अब तक ये समस्त तत्व हम सबका जीवन सुरक्षित रखे हुए हैं। बिना किसी बैरभाव के इनका एक ही लक्ष्य है, हमारे अस्तित्व को बचाए रखना। बदले में हमने इन्हें क्या दिया, यह सोचने मात्र से ही हमारा हृदय अपराध-बोध से भर जाता है।

खैर! मनुष्य को तो जीवन मूल्य भी सीखने थे परन्तु वह प्रकृति से कोई ज्ञान नहीं ले सका। स्वभावतः उस पर अपना अधिकार ही समझ बैठा! इन्हीं नासमझ मानवों के लिए श्रीकृष्ण अवतरित हुए। मुझ जैसा सतही तौर पर धार्मिक जानकारी रखने वाला इंसान भी इतना तो समझता ही है कि उन्होंने जन-जन को प्रेम के मायने सिखाए। मित्रता कैसी होती है, यह बतलाया। श्रीकृष्ण ने न केवल स्त्रियों की कोमल भावनाओं को महत्त्व दिया बल्कि उन्हें सशक्त करने में भी अपार योगदान दिया। उन्हें सम्मानपूर्वक देखा, उनकी भक्ति का मान रखा। विष को अमृत में बदल दिया। करुणा के भाव का जनमानस में संचार किया। धनवान और निर्धन के मध्य कोई अंतर न रखा और सारे उदाहरणों को प्रस्तुत कर मानव जाति को भेदभाव से ऊपर उठ जीवन जीने का सन्देश दिया।

कृष्ण, स्त्रियों को इसीलिए सर्वाधिक प्रिय हैं क्योंकि वे उन सा साथी चाहती हैं, जो प्रेम करे तो ऐसा जो उन्होंने राधा से किया। वो दोस्त बने तो गोपियों के सखा जैसा या फिर ऐसा जो सुदामा के प्रति था। भाई हो तो ऐसा जो सुभद्रा ने पाया। एक रक्षक हो तो

ऐसा जो द्रौपदी के लिए था। नायक हो तो ऐसा जिसके होठों पर मुरली सज प्रेम धुन गुनगुनाती है और सब उसकी ओर खिंचे चले आते हैं। एक ऐसा प्यारा साथी जो अवसर मिलते ही मीठी शैतानियों से भी बाज नहीं आता!

कृष्ण के हर रूप में प्रेम है। उनके व्यवहार में प्रेम है। उनकी आँखों से प्रेम घटा बरसती है और उनकी मुरली जब कोई मीठी धुन छेड़ती है तो सम्पूर्ण वातावरण प्रेममय हो जाता है। राधा-कृष्ण का प्रेम अमर है तो इसलिए क्योंकि उन्हें एक दूसरे का सम्मान करना आता था, परस्पर भावनाओं की कद्र थी, प्रेम इतना कि दूर होकर भी रत्ती भर कम न हुआ। दोनों एक दूसरे के बिना अधूरे थे। प्रेम वही सच्चा होता है जो अनकंडीशनल हो, जिसमें हम दूसरे पक्ष को जस का तस स्वीकार करें, जिसमें आकर्षण से परे आत्माओं का मिलन पहली प्राथमिकता हो। यही वह प्रेम है जो बिछोह के बाद भी रहता है। युगों-युगों तक रहता है।

जब ईश्वर ने हमें प्रेम के हर रूप से मिलवाया है तो उसके पूजक ईर्ष्या, वैमनस्य, अपराध कहाँ से सीख गए? हम जिस मुरलीधर का गुणगान करते नहीं थकते, जिसका रूप हमें मोह लेता है। उसके जीवन मूल्यों को क्यों नहीं अपना पाते? हम तो प्रेम के दुश्मन अधिक बन बैठे हैं। परिवारों को खत्म कर देने को तैयार हैं! हमें प्रेम कहानियाँ पसंद तो हैं पर फिल्मी! असल जीवन में तो तलवारें खिंच जाती हैं, इज़्ज़त दाँव पर लगती नजर आती है, समाज में नाक कट जाने का भय खाने दौड़ता है।

जो हम कृष्ण के सच्चे साधक हैं तो हमें उन्हें पूजने के साथ ही, इस वर्तमान समाज का प्रहरी मान उनसे सीख लेनी चाहिए। श्रीकृष्ण का सांस्कृतिक और सामाजिक स्वरूप हमारी संस्कृति और सभ्यता के लिए आज भी उतना ही प्रासंगिक है जितना सदियों पूर्व रहा होगा।

खग ही जाने, खग की भाषा

बीते माह एक खबर चर्चा में रही जिसमें संन्यास जीवन को अपनाने के बाद दो लोगों ने विवाह का निश्चय किया और अब वे गृहस्थ जीवन में प्रवेश कर चुके हैं। कारण था मोहब्बत! विवाह की इससे अधिक खूबसूरत वजह कुछ और होनी भी नहीं चाहिए! प्रेमियों का हक है ये कि वे उम्र भर साथ रहें! प्रेम में डूबे लोग, किसी का दिल नहीं दुखाते, किसी को लूटते नहीं, परेशान नहीं करते। वे प्रेम से रहना चाहते हैं और इसी भाव की अपेक्षा दूसरों से भी रखते हैं।

अच्छी बात यह है कि इस शादी से वहाँ उपस्थित लोग भी प्रसन्न होते नजर आ रहे थे और किसी ने भी कोई ऊलजलूल बात नहीं की बल्कि कई संन्यासी इस शादी के गवाह बने हैं। उस वीडियो को देखने-सुनने के बाद लगता है कि समाज के एक विशेष टेढ़े वर्ग को इन गाँव वालों से सीख लेनी चाहिए! ये वर्ग जो सोशल मीडिया पर अपने आपको कूल और अल्ट्रा मॉडर्न समझते नहीं थकता लेकिन अमृता सिंह से लेकर प्रियंका चोपड़ा और मलाइका अरोड़ा तक को अपने पति की माँ बोलने से नहीं चूकता! उन पर छींटाकशी और भद्दे मजाक करता है! उस पढ़े लिखे, नफरत पोसते वर्ग को बटेश्वर के इन लोगों से यह ज्ञान प्राप्त कर ही लेना चाहिए कि 'खग ही जाने खग की भाषा'। प्रेम का अर्थ केवल प्रेम ही होता है और यह उसके अलावा और किसी भी भाषा में यकीन नहीं करता। प्यार में उम्र तो कोई मायने ही नहीं रखती!

'न उम्र की सीमा हो, न जन्म का हो बंधन

जब प्यार करे कोई, तो देखे केवल मन

नयी रीत चलाकर तुम, ये रीत अमर कर दो'

इंदीवर के लिखे इस गीत को जगजीत सिंह ने हमेशा-हमेशा के लिए अमर कर दिया है लेकिन दुर्भाग्य की बात यह है कि लगभग 40 साल बाद भी यह रीत अब तक इस समाज में सहजता से नहीं स्वीकारी गई है। बल्कि समाज तो शब्दशः इसका उल्टा करके चलता है। वो उम्र देखता है, जन्म, जाति, पैसा सब में भरपूर दखल देता है पर मन देखना और उसे समझना इसने सीखा ही नहीं! दुनिया भर को प्रेम का पाठ पढ़ाने वाले हमारे देश में सबसे ज्यादा गाज प्रेमियों पर ही गिरती आई है। सबसे अधिक प्रश्न प्रेमियों से ही पूछे गए हैं, सबसे अधिक तिरछी निगाहों का सामना उन्हें ही करना पड़ा है। सबसे अधिक हतोत्साहित (जलील ही पढ़िए इसे) उन्हें ही किया गया है, सबसे अधिक नफरत और बद्दुआ उन्हीं के पवित्र हृदय को पीड़ा देती आई है। धर्म और इज्जत के नाम पर हिंसा का खूनी खेल, सबसे अधिक प्रेमी युगल के साथ ही खेला जाता है।

हम जिस 'भद्र' समाज का हिस्सा हैं उसी में आज भी स्त्री की उम्र अधिक होने पर, पुरुष को फँसाने जैसे घिनौने इल्जाम भी उसी के मत्थे मढ़ दिए जाते हैं। अक्सर ही प्रेम को किसी एक पक्ष की साजिश कहकर, उसके खिलाफ नई किस्म की साजिश रच दी जाती है। इस तरह के लोगों का लक्ष्य एक ही है, प्रेम को नष्ट करके, घृणा के बीज बोना और फिर ये उम्मीद भी करना कि इसका मैडल भी उनको नवाजा जाए कि कैसे वे इस समाज के कट्टर रक्षक बन इस दुनिया में अवतरित हुए हैं।

बटेश्वर धाम के इस जोड़े का विवाह, निम्न मानसिकता से भरे लोगों की सोच के सामने एक सहज उत्तर बनकर प्रस्तुत होता है। ये हारे-घबराए हुए लोगों को उम्मीद का उजाला भी देता है। साथ ही समाज को ये उदाहरण भी देता है कि जीवन के किसी भी दौर में जब अकेलापन महसूस हो, बेहिचक उस अपने का हाथ थाम लो जिसे हमारे लिए ही बनाया गया है।

ईश्वर करे कि उसके दरबार में फलते-फूलते इस प्रेम को, विशेष आशीर्वाद प्राप्त हो। इस युगल का वैवाहिक जीवन इसी प्रेम की छाया में पल्लवित होता रहे और हमें प्रेम की इतनी कहानियाँ पढ़ने को मिलें कि हमारा मन अचरज से न भरे!

अब समय आ चुका है कि प्रेम इस समाज का अभिन्न भाव बनकर रहे, उसमें भीतर तक समा जाए। प्यार, इश्क, मोहब्बत की कहानियाँ इतनी आम हो जानी चाहिए कि इन्हें खास समझ, इस पर चर्चा बंद ही हो जाए।

सोशल मीडिया वाली 'भक्ति' और बिखरते रिश्ते

इन दिनों, सोशल मीडिया प्लेटफ़ॉर्म लोगों को धीरे-धीरे डसने लगे हैं। 'उसकी शर्ट मेरी शर्ट से सफ़ेद कैसे' के दिन अब लद चुके हैं। बीते एक दशक से अब इसके स्थान पर 'उसके लाइक, कमेंट, फॉलोअर मेरे से अधिक कैसे' ट्रेंड कर रहा है। प्रेमी-प्रेमिका से नाराज हो उठता है कि 'मेरी पोस्ट छोड़ तूने उसकी पोस्ट पर दिल क्यूँ बनाया!' शक को दूर करने के लिए पासवर्ड भी रख लेता है। कई पति भी अपनी पत्नी का पासवर्ड रखते हैं कि 'ये तो भोली है, मुझे ही इसका ध्यान रखना पड़ेगा'। जबकि उसका उद्देश्य नजर रखने का रहता है। कहीं मित्र, रिश्तेदार भी मुँह फुलाये घूमते कि 'देखो! कितने भाव बढ़ गए, हम भी उसकी पोस्ट पर नहीं जाएंगे अब!' ऐसे कई किस्से आप रोजमर्रा सुनते ही होंगे। ईर्ष्या, प्रतिस्पर्धा का ये आलम तब है जबकि इससे कोई आर्थिक लाभ ही नहीं! कोई न गिनने वाला!

बस मन को संतुष्टि मिलती है कि इतने लोग हमें जानते हैं। चाल में एक ठसका आ जाता है। मुस्कान डेढ़ के बजाय दो इंच हो जाती है। अब उसके बाद क्या? वही तुम वही हम! जब मुसीबत में होंगे तो इनमें से कोई भी न आएगा। गिने-चुने करीबियों को छोड़ दिया जाए तो ये हजारों, लाखों की संख्या अचानक ही एक बड़े खोखले शून्य में परिवर्तित होती दिखेगी। आप फिर एक बार वहीं खड़े होंगे, जहाँ से ये सफर शुरू किया था!

बीते सालों से एक नया नाटक और प्रारम्भ हो चुका है। वो है भक्त और अभक्त में बहस का। बहस तो अत्यंत शालीन शब्द है, असल में तो बात जूतमपैजार तक पहुँचती है। अब यदि कोई सच में अपने आराध्य की भक्ति की बात करे, तब भी सामने वाले की नजर में मोदी जी का चेहरा ही चमक उठेगा। ये ईश्वर की महिमा समझें या माननीय का प्रताप, ये आप पर निर्भर है। आप किसी को फेंकू कहें तो वह तुरंत आपको पप्पू कह देगा। संकेत दोनों पक्ष समझते हैं। देश की डिक्शनरी बिना ऑक्सफोर्ड हस्तक्षेप के, अपने-आप ही बदल चुकी है। 'भक्त' का अर्थ बना 'किसी की धुन में इतना अंधा हो जाना कि आँखों के सामने रखे सच को भी अपने कुतर्कों से नकारने लगें'। कुछ कट्टरों ने तुरंत एक तोड़ निकाला और 'सेक्युलर' को गाली की तरह परोस दिया। इसी दौरान 'देशद्रोही' शब्द का भी पुनर्जन्म हुआ और एक नई किस्म की नफरत अँगड़ाई लेने लगी। देश स्वतः ही दो खेमों में बँट गया। कल जो मित्र साथ हँसते-बोलते थे, वे अब अपने-अपने पालों में खड़े होकर नित हर शब्द का अनुवाद बनाते हैं, गोया हमारी रसोई मोदी जी या राहुल जी ही चला रहे।

पति-पत्नी की लड़ाई के बीच अब बच्चे या घर-खर्च नहीं आता, मोदी जी आते हैं! क्योंकि एक समर्थक और दूसरा घोर विरोधी। अब बताइए, जिसके लिए लड़ रहे उसे तो पता तक नहीं और इधर आपने रिश्ते ही दाँव पर लगा दिए। रिश्तेदारों के मुँह भी आपको देख टेढ़े होने लगे हैं या वो खुन्नस भरी नजर से देखते हैं। काहे को? क्योंकि तुम्हें वो पसंद हैं और इन्हें कोई और। इतना तो कॉलेज के लड़के भी अपनी प्रेमिका के लिए नहीं लड़ते होंगे जितना युद्ध आजकल लोगों की पोस्ट पर दिख जाता है। देश में विकास की ये लहर सचमुच अजीब और खतरनाक है। ऐसा राजनीतिकरण सदियों बाद देखने को मिला है जहाँ आम लोग बेवजह भिड़ रहे।

लोगों को राजनीति जैसे नीरस विषय में अचानक ही दिलचस्पी होने लगी है, ऐसा भी नहीं! दरअसल सोशल मीडिया की अंधाधुंध दौड़ में हमारे पास कोई विकल्प ही नहीं बचा, किसी एक को चुनने के सिवाय। अगर चुना नहीं तो आप ढोंगी कहलाये जाएंगे। व्हाट्स एप्प, टीवी, रेडियो, अखबार सब जगह इन्हीं के जोक्स...! अब इंसान हैं तो कभी-कभार हँसी निकल भी जाती है। बस, इधर आप हँसे, उधर आप पर भाजपाई या कांग्रेसी होने का ठप्पा लग गया। विरोधी हो तभी तो फॉरवर्ड किया तुमने! ऐसे-कैसे हँस सकते हो जी! पिछली बार जब हमने चुटकुला सुनाया, तब तो ऐसी खींसें नहीं निपोरी थीं तुमने? ब्लाह, ब्लाह, ब्लाह! आप सिर पीटते रहें (अपना ही) पर आक्षेप का ये दौर अंतहीन ही रहेगा। आप कोई भी एक वाक्य बोलें, उसे किसी एक खेमे का ही माना जाएगा।

इस सबमें वे लोग टूटते रिश्ते या खुद अपनी ही मौत मारे जाते हैं जिन्हें केवल 'देश' से मतलब। जो किसी भी किस्म की कट्टरता को देश पर हावी नहीं होने देना चाहते, जो 'हत्या' में धर्म नहीं टटोलते, जिनके लिए 'एक भारत' ही 'श्रेष्ठ भारत' है। ये इस देश का दुर्भाग्य है कि इन चंद लोगों पर ही सबसे ज्यादा प्रश्न उठते हैं। और वे हर वक़्त अपनी सफाई देते नजर आते हैं। उन्हें बिन पेंदी का लोटा, धर्मनिरपेक्षता की पूँछ और कई निम्नस्तरीय अपशब्दों से भी नवाजा जाता है। भारत के इस रूप की कल्पना हमने कभी नहीं की थी। आखिर ये राजनीतिक दल हमें भिड़ाकर अपना उल्लू सीधा करने के सिवाय और कर ही क्या रहे हैं? उस पर दिक्कत ये कि लोग उल्लू बनने को नतमस्तक हैं। उन्हें अपने आका को खुश करके ईनाम जो लूटना है! आज लूट लीजिए और बाद में अपने भाग्य पर बुक्का फाड़ रोइएगा। पर उससे पहले एक सूची भी बनाते चलिए कि इन निर्मोही, स्वार्थी नेताओं के समर्थन के चक्कर में आपने कितने अपने खो दिए हैं। अब इसकी भरपाई कौन करेगा? ये लॉन्ग डिस्टेंस से वोट बैंक भरने वाले निष्ठुर आका? जिन्हें आपका नाम तक पता नहीं!

मूल्यांकन का मूल्य

बारहवीं की परीक्षाओं के निरस्त होने के बाद से, विद्यार्थियों के मूल्यांकन को लेकर उठे प्रश्नों पर विराम लग गया था। तय किया गया था कि दसवीं, ग्यारहवीं के तीस-तीस प्रतिशत और बारहवीं प्री बोर्ड/आंतरिक परीक्षाओं के चालीस प्रतिशत को लेकर परिणाम तैयार किये जाएंगे। जैसा कि होना ही था, इस निर्णय पर कई लोगों ने आपत्ति दर्ज की।

प्रायः यह देखा गया है कि जिन विद्यार्थियों के दसवीं में कम अंक आते हैं उनमें से कुछ इतनी मेहनत करते हैं कि बारहवीं में कमाल कर जाते हैं। वहीं इस तथ्य को भी झुठलाया नहीं जा सकता कि ग्यारहवीं में विद्यालय द्वारा बड़ी कठोरता से मूल्यांकन किया जाता है जिससे छात्र बोर्ड परीक्षाओं को अधिक गंभीरता से लें! अब जिन्होंने पूरे वर्ष अथक परिश्रम किया, उन्हें इस निर्णय से निराशा ही हाथ लगी। यूं इस बात से भी इंकार नहीं किया जा सकता कि बोर्ड चाहे जो तरीका चुनता, उसे सभी विद्यार्थियों की सहमति मिल पाना असंभव ही होता! विकल्प से मूल परिणाम जैसी संतुष्टि पा लेना उतना सरल नहीं है!

अब जैसा बोर्ड परीक्षा परिणाम के लिए तय हुआ था कि टॉप तीन विषयों के एवरेज को ही पूरा रिजल्ट मान लिया जाएगा लेकिन क्या यह कमजोर विषयों से मुँह चुराना न हुआ? पुराने प्रदर्शन के आधार पर या उनका औसत निकालकर आज का निर्णय ले लेना कितना सही है? इस पर चर्चाएं होती रहेंगीं। लेकिन सामाजिक एवं व्यक्तिगत जीवन में भी तो हमारी सोच इससे अलग कहाँ है?

जीवन के किसी भी क्षेत्र में जब किसी का मूल्यांकन करना होता है तब हम इसी सूत्र को तो अपनाते हैं। अंतर इतना ही है कि हम उस व्यक्ति के अच्छे-बुरे कार्यों को अपनी सुविधानुसार वरीयता देकर उसके बारे में एक दृष्टिकोण तय करते हैं और फिर उसे अच्छे या बुरे इंसान की श्रेणी में डाल देते हैं। यदि किसी व्यक्ति ने हमारी कोई सहायता की है तो वह भला लगता है और नहीं की, तो अचानक से निकृष्ट लगने लगता है। कोई प्रशंसा करे तो उससे प्यारा कोई नहीं और आलोचना करने वाला सदा खटकता आया है। यही कारण है कि सच्चे और स्पष्टवादी व्यक्ति को पसंद करने वाले बहुत कम होते हैं। हमारे समाज में चापलूसी का बोलबाला यूं ही नहीं हो गया है!

हम स्वयं भी तो ऐसे ही हैं। यही अच्छा एवं सुविधाजनक लगता रहा है हमें। हम अपने गुणों को पसंद करते हैं। हमारी

खूबियों को दुनिया के सामने प्रदर्शित करना चाहते हैं लेकिन हममें जो कमियाँ हैं उन्हें ढक लिया करते हैं। उन्हें सुधारने का कोई यत्न नहीं करते! सामाजिक ढांचे द्वारा पोषित इस परंपरा का प्रशिक्षण बचपन से ही प्रारंभ हो जाता है। बच्चा किसी विषय में अधिक अंक ला रहा है और किसी में कम। तो हमने मान लिया कि जिसमें अधिक अंक हैं, उस विषय में उसकी रुचि है तथा जिसमें अच्छे अंक नहीं आ रहे, वह उसे समझने में कमजोर है। यहाँ यह तथ्य विचारणीय होता ही नहीं कि यदि उसकी बुद्धि एक विषय में अच्छी चल रही, तो दूसरे में क्यों नहीं? क्या ये नहीं हो सकता कि उक्त विषय के अध्यापक ही सही से नहीं पढ़ा रहे हों! लेकिन नहीं, कमजोरियों को ठीक करने या उसका कारण पता करने के स्थान पर, बच्चे के मस्तिष्क में यह बात पहुँचा दी जाती है कि 'सुनो, जिसमें अंक कम आ रहे हैं, उसमें तुम्हारी रुचि नहीं!'

तत्पश्चात इसी के आधार पर छात्र का भविष्य तय होता है। अधिकतम अंकों के आधार पर कोई विज्ञान में अच्छा, कोई वाणिज्य में, तो कोई कला में, और वही विषय उसने ग्यारहवीं में चुन लिया। उसे इस बात की प्रसन्नता रही कि कमजोर विषयों से पीछा छूटा! फिर बारहवीं में भी यही स्थिति सामने आई और उसने ठीक वही प्रक्रिया दोहराई। उसने कमजोर अंक वाले विषयों को छोड़कर, अधिक अंक वाले तीन विषयों पर फोकस किया। इस बार इस चयन में वही नहीं, वे सभी कॉलेज भी शामिल हुए जो अंकों के आधार पर प्रवेश देते हैं। छात्र ग्रेजुएट हो गया। अब पी.जी भी इसी सूत्र से हुआ। अबकी बार केवल एक विषय के साथ आगे बढ़ना था। उसने सर्वाधिक अंकों वाला चुन लिया और जहाँ कमी थी उसे झटकने का प्रोत्साहन एवं प्रावधान तो अपनी शिक्षा-प्रणाली में पहले से है ही!

जी, हाँ! पूरे विद्यार्थी जीवन में हम अपनी कमजोरियों से लड़ने के बजाय उसे नजरअंदाज कर आगे बढ़ते रहे क्योंकि हमारी शिक्षा-व्यवस्था ही ऐसी है जो अंक देखती है। रुचि या बुद्धि से उसका कोई सरोकार नहीं होता! फिर आगे जाकर यही विद्यार्थी जब हताश होते हैं तो अपनी पसंद की ओर लौटने लगते हैं। कोई फोटोग्राफी चुनता है, कोई चित्रकारी, कोई लेखन तो कोई अपने बचपन के गीत-संगीत, नृत्य के शौक को पूरा करने के लिए क्लासेस शुरू करता है। कोई निजी व्यवसाय प्रारंभ कर देता है। लेकिन उन वर्षों का हिसाब कौन देगा जो अरुचि के साथ गुजर चुके हैं?

कलाकार हिट तो अच्छे, फ्लॉप तो बेकार। लेकिन प्रतिभा का क्या? वे जिन्होंने अपनी फ्लॉप फिल्मों से सबक लिया और अपनी प्रतिभा को निखारा, वे लंबे समय तक लोकप्रिय बने रहे। जो एक ही ढर्रे पर चलते रहे, दर्शकों की निगाहों से शीघ्र ही उतर भी गए।

सरकारों को देखिए, वे भी अपना मूल्यांकन अपने अच्छे कामों की वजह से ही करवाना चाहती हैं। कमियों को दिखाते ही आप पर तरह-तरह के आरोप जड़ दिए जाते हैं। ये हम मनुष्यों की आम मानसिकता है कि गुण गिने जाएं, दोष पर बात बाद में होगी! सरकार हो, या हम... हम हमेशा अपनी ताकत के भरोसे आगे बढ़ना चाहते हैं। और उस ताकत में कोई खामी आती है तो उसे भी पीछे छोड़कर आगे बढ़ने की कोशिश करते हैं। आगे बढ़ने के लिए ये ठीक भी है लेकिन, क्या इससे एक आशंका यह नहीं रह जाती है कि कमजोरियों को पीछे छोड़ते-छोड़ते एक दिन जब सारी ताकत ही खत्म हो जाएगी, फिर क्या करेंगे हम? हमने कमजोरियों को बेहतर किया नहीं और जिस एक में शक्ति थी वो सारी चुक गई! अब क्या??

हमने छः विषय पढ़ना शुरू किए। तीन खराब लगे, छोड़ दिए। फिर तीन में से भी दो खराब लगे, वे भी छोड़ दिए। बाद में पता चला कि वो जो एक था, उसमें भी कुछ खास मजा नहीं आया। फिर अपना लक्ष्य ढूँढने को भटकने लगे। इसके बाद अवसाद, कुंठा और परिवार पर आरोप-प्रत्यारोप का बुरा दौर प्रारंभ होता है। कि इसके कारण ये नहीं कर सके, वो नहीं हो सका! पर इस सबका हासिल क्या? अब वह बीता समय कौन लौटा सकेगा?

सोचना होगा कि मूल्यांकन का मूल्य क्या है?

उसका उद्देश्य क्या है?

क्या रिजल्ट सिर्फ मूक दर्पण होता है, या वो कभी हमसे बात भी करता है?

याद है हर रिजल्ट के नीचे क्लास टीचर का एक नोट आता था–

किसी को कहा जाता था, 'आपको इस विषय में थोड़ा और ध्यान देने की जरूरत है (यू नीड टू फोकस मोर)'।

तो किसी को कहा गया कि 'कॉन्सेप्ट इज क्लियर बट यू मस्ट इंप्रूव योर हैंडराइटिंग'।

किसी ने ये सुझाव भी दिए 'की वर्ड्स को न भूलें!',

'हड़बड़ाएं नहीं!', 'अनुशासन में रहें', 'इस प्रश्न का उत्तर अंत में लिखा है, कृपया अगली बार ऐसा न लिखें एवं व्यवस्थित तरीके से क्रमवार उत्तर दें'।

जिसने भी इन सुझावों पर अमल किया वो तर गया क्योंकि वो शब्द ही मूल्यांकन का मूल्य है।

आखिर क्यों अनुत्तीर्ण हुए हिन्दी के विद्यार्थी?

बीते दिनों उ.प्र. माध्यमिक शिक्षा बोर्ड द्वारा हाईस्कूल और इंटरमीडिएट परीक्षा परिणाम में लगभग 8 लाख परीक्षार्थी हिन्दी विषय में फेल हो गए। उधर परिणाम आए और इधर उनकी स्वाभाविक विवेचना प्रारम्भ हुई। इन लम्बी-चौड़ी बहसों में अधिकांश का जोर इसी बिंदु पर था कि हिन्दी भाषी राज्य होते हुए भी भाषा की ये दुर्गति!

प्रश्न उठता है कि हिन्दी की दुर्दशा कहाँ नहीं हैं? किस प्रदेश में उसे सम्मान की दृष्टि से देखा जाता है?

यह दयनीय स्थिति क्यों है?

जब आप संभ्रांत मनुष्यों की भाषा का मुकुट आंग्ल भाषा को पहनाते हैं, उसी समय आप अपनी भाषा का दुखद भविष्य निश्चित कर देते हैं।

घर में अतिथियों के आने पर बच्चों का ज्ञान प्रदर्शन करवाते समय उनसे अंग्रेजी भाषा की लघु कविताएँ सुनवा, माता-पिता बलिहारी हुए जाते हैं परन्तु इन बच्चों को कहरा सिखाने का कष्ट कोई नहीं उठाता! उन्हें स्वयं याद होगा, यह तथ्य भी संदेह के घेरे में ही समझिये। दोष इन दोनों का नहीं अपितु उस शिक्षा व्यवस्था का है जो प्रतिदिन हमें यह स्मरण कराती है कि आंग्ल भाषा ही आपका उज्ज्वल भविष्य निर्धारित करेगी।

प्रतियोगी परीक्षाओं में भी यही प्रधान भाषा के रूप में दर्शित होती है।

किसी के सामने अंग्रेजी में बात करने का अलग ही रौब समझा जाता है। तो बच्चा क्यों नहीं, हिन्दी छोड़ इस भाषा को प्रधानता देगा! उसे भी तो इसी समाज का हिस्सा बनना है। अपने भविष्य को सुनहरा होता हुआ वह भी देखना चाहता है। ज्ञातव्य हो, यह सोच हमने ही उसे दी है।

कमी यहाँ रह जाती है-

भूल यह भी हो जाती है कि हम अपनों को बड़े ही हल्के में ले लेते हैं। 'अरे! तुम तो अपनी ही हो! तुम तो समझोगी ही!' या कि 'ये तो घर की बात है, बाद में देख लेंगे!' जब हम रिश्तों को इस मानसिकता के साथ जीने लगे हैं तो भाषा की बिसात ही क्या! बड़े भी कहते हैं कि 'शेष विषयों पर ध्यान दो, हिन्दी में क्या सीखना! ये तो रोज की भाषा है।' बच्चे जिन्हें पहले से ही इसे पढ़ना 'कूल' नहीं लगता, वे भी यह मान बैठते हैं कि 'रोज तो बोलते हैं। इसमें क्या! उत्तीर्ण तो हो ही जाना है।' उनका सारा ध्यान अन्य विषयों पर केंद्रित हो जाता है। यही अति आत्मविश्वास उन्हें ले डूबता है।

वे मौखिक और लिखित भाषा का मूल अंतर भूल जाते हैं। व्याकरण की अल्पज्ञता का प्रभाव उनकी अंकसूची पर चिह्नित होता है।

दोष किसका है?

अब इसमें आश्चर्यचकित होने जैसा क्या है? एक तो बच्चे को पहले से ही हिन्दी में रुचि जागृत नहीं हुई, उस पर इस विषय की उपयोगिता भी हम सार्थक सिद्ध नहीं कर सके! ऐसे में उसका झुकाव अन्य विषयों की ओर होना तय ही है। हमारा सामाजिक ढाँचा और शैक्षणिक नीतियाँ ही ऐसी हैंय जहाँ अंग्रेजी को, हिन्दी से अधिक सम्मान हर क्षेत्र में दिया जाता है। फर्राटेदार अंग्रेजी न बोलने वाले प्रायः उपेक्षित ही होते रहे और हीन भावना का शिकार हुए। दोष उस व्यवस्था का है जिसने अपनी ही भाषा का अपमान कर दूसरी के सिर पर ताज रख दिया। अंग्रेजी से कोई आपत्ति नहीं परन्तु प्रश्न यह है कि उसके सामने हिन्दी कमतर क्यों आँकी जाती है?

वर्तमान स्थिति क्या है?

ऐसा नहीं कि हिन्दी की प्रिंट पत्र-पत्रिकाएँ उपलब्ध नहीं सोशल मीडिया पर भी इनकी संख्या पर्याप्त है लेकिन वर्तमान पीढ़ी में इस भाषा के प्रति रुझान न्यूनतम ही देखने को मिलता है। कारण ऊपर उल्लिखित किये ही जा चुके हैं।

मानकर चलिए कि बच्चों में जो थोड़ी-बहुत हिन्दी शेष है, उसके लिए बॉलीवुड फिल्मों और गीतों का बड़ा योगदान है।

हिन्दी सेवा के नाम पर प्रतिवर्ष संगोष्ठी, परिचर्चा इत्यादि कार्यक्रमों के माध्यम से लाखों फूँका जाता है पर हिन्दी की स्थिति जस-की-तस बनी हुई है। कारण यही कि कागजी भाषणों में बड़ी-बड़ी बातें कह दी जाती हैं, तालियाँ पिटती हैं, रंगीन तस्वीरें सुर्खियों के साथ प्रकाशित होती हैं। बस, सेवा यहीं समाप्त हो जाती है। ज़मीनी तौर पर कोई कारगर नीति भूले से भी दृष्टिगोचर नहीं होती!

भाषा, हमारी संस्कृति का अहम हिस्सा है, अब इस पर विशेष ध्यान देने की आवश्यकता आन पड़ी है।

सफलता को मारिये गोली, पहले असफल होना सीखिए!

हमारे बच्चों को हर हाल में ये समझना होगा कि परिवार के लिए उनसे ज्यादा जरुरी कुछ नहीं होता! ये नाम, पैसा, शोहरत किसी के साथ नहीं टिकते, कभी नहीं टिकते फिर इनके पीछे इस हद तक क्यों भागना कि एक दिन इनके बिना जीना ही दूभर लगने लगे! यदि आप अपनी समस्या बताएंगे नहीं, तो किसी को पता कैसे चलेगा? घर, परिवार, मित्र में से कोई-न-कोई तो अवश्य ही सुनता है, हमेशा सुनता है। ये मानती हूँ कि बहुधा हमारे पास किसी की समस्या का समाधान नहीं होता लेकिन उससे भी कहीं ज्यादा ये विश्वास रखती हूँ कि कह देने से मनों बोझ हट जाता है दिल से। दुःख की परतें थोड़ी झीनी होने लगती हैं। जीवन उतना कठिन नहीं लगता कि जिया ही न जा सके! ऐसे कैसे आप किसी भी समस्या को जीवन से बड़ा मान लेते हो कि वो न सुलझी तो सब कुछ बेकार है? खत्म हो गया है!

जब आप इस दुनिया में अपनी मर्जी से आए नहीं तो आपको कोई हक नहीं कि जाने का रास्ता स्वयं चुनें। वृक्ष में फूल आते हैं, सूख जाते हैं। फल बनते हैं, खा लिए जाते हैं। तो क्या वृक्ष फलना-फूलना बंद कर दे? वनस्पति तो फिर उगे ही न कभी! प्रकृति में जितनी खूबसूरत चीजें हैं, सदियों से सब अपनी जगह हैं। हम मनुष्यों ने इनका सीना चीर दिया है पर वे तब भी हैं...अडिग! पता है क्यों? क्योंकि इन्हें किसी से कोई अपेक्षा नहीं होती। इन्होंने देना ही सीखा है, इसीलिए निराशा इनके पास फटकने के बहुत पहले उलटे पाँव लौट जाना ही पसंद करती है। आप वृक्ष बनना सीखिए। सपने देखना अच्छी बात है। उसे पूरा करने की कोशिश, उससे भी कहीं अधिक अच्छा। लेकिन जो ये सपने पूरे न हुए तो बिखरना क्यों है? एक कोशिश और नहीं हो सकती क्या? भूलिए मत कि आप भी किसी का सपना हो सकते हैं।

कामयाबी की कहानी उन पुराने लोगों से सुनिए, जिन्होंने अपना घर बनाने में ही पूरा जीवन निकाल दिया। जिन्होंने कभी एसी वाला अपना कमरा या हवाई जहाज की यात्रा का ख्वाब तक नहीं देखा। वो इसलिए क्योंकि उन्हें अपने पैरों और चादर का माप पता था। उनके लिए आज भी उस वस्तु की अहमियत है जो उन्होंने पाई-पाई जोड़कर खरीदी थी। फिर चाहे वो पुराना रेडियो हो या फ्रिज। आपको भी हर सुविधा की कीमत और अहमियत समझनी होगी। रिश्तों का मूल्य समझना होगा। उन लोगों की भावनाओं की कद्र करनी होगी जो आपसे हृदय से जुड़े हुए हैं। मौत को गले

लगाने का निर्णय लेने से पहले एक बार उन्हें गले लगाइए। कुछ भी न महसूस हो तो बात कीजिए उनसे। हद है! ऐसे कैसे सबको छोडकर यकायक चले जाते हैं लोग।

अरे, जिसका जितना जीवन है वो जीता है। परेशानियाँ, संघर्ष, दुःख किसके साथ नहीं चलते? इन्हीं से जूझते, लड़ते, कभी गिरते, कभी उठते, संभलकर चलने का नाम ही तो है जिंदगी। इसमें घबराने जैसा क्या है? हिम्मत है तो जीकर दिखाइए और नहीं है तो... ! सीखिए जीना!

सफलता को मारिये गोली, पहले असफल होना सीखिए। जीने की राह यहीं से निकलती है। चखिए, हारने का स्वाद! पीजिये खून के घूँट! जी करे तो बाल्टी भर रो लीजिए। मन लगाने के लिए संगीत सुनिए, बागवानी कीजिए, पढ़िए-लिखिए। कुछ भी कीजिए, जो आपको कभी पसंद हुआ करता था पर प्लीज मरने का ख्याल भी न लाइए। जीवटता बनाए रखिए।

'आत्महत्या' धोखा है, सारे रिश्तों के प्रति!

'जीवन जीना शुरू कर दिया, फिर हारना नहीं है!'

लॉकडाउन के बाद, हम कितना बदले?

लॉकडाउन में एक लम्बा समय बिताने के बाद कई महत्वपूर्ण प्रश्न सामने आ खड़े हुए और मैं यह सोचने पर विवश हो गई कि उस समय से हमने क्या सीखा? हमने जीवन को जिस तरह जीना सत्य मान लिया था, क्या वह सही था? उतना ही था? ऐसे कई प्रश्न इस समय हृदय को उद्वेलित कर रहे हैं।

पहली बात जो मस्तिष्क को झिंझोड़ रही है वह ये कि क्या अब हम पहले से बेहतर मनुष्य हो गए हैं? तो जवाब मिलता है कि किसी के भी व्यक्तित्व में दृष्टमान परिवर्तन इतनी आसानी से नहीं होते! एक ही झटके में सारे बुरे लोग अच्छे नहीं हो सकते और न ही सबमें सामूहिक मानवता जाग सकती है। ईर्ष्या-द्वेष, घृणा की राजनीति और समाज में दुर्भावना फैलाने वाले लोगों में भी सेंसेक्स की तरह भीषण गिरावट नहीं आई है लेकिन जो थोड़े कम बुरे हैं या कुंठा ने जिनकी अच्छाइयों को अपने घेरे में ले लिया थाय उन्होंने विनम्र होना अवश्य सीख लिया है। उनकी आवाज की तल्खी विलुप्तप्राय है। अब वे डॉक्टर्स को लूटने वाला नहीं कहते बल्कि उनकी कहानियाँ कहते-सुनते आँख नम कर लेते हैं। हमें छोटी-बड़ी वस्तुओं की कद्र करना आ गया है। वो सब्जी वाला जिससे मुफ्त में धनिया लेने और सब्जियों को सही तौलकर देने के लिए चिकचिक हुआ करती थी अब उसकी एक पुकार पर मन कृतज्ञ हो उठता है कि इतने मुश्किल समय में भी उसने हम सबकी गलियों से गुजरने में कोई आपत्ति नहीं दर्ज की! साथ ही उसी मुस्कान के साथ यह ढांढस बंधाना नहीं भूलता कि 'बहनजी, कल फिर आऊंगा।' वो स्वीपर अब भी उसी तन्मयता के साथ सड़क पर झाड़ू फेर जाती है और कचरे वाला भी बिना नागा आकर गंदगी समेट लेता है। गर ये न आते तो सोचो किस तरह रहा जाता, हमारे अपने ही घरों में! इनका नियमित आना हमारी संवेदनाओं को पुनर्जीवित करने जैसा है।

दूसरा प्रश्न जीवन मूल्यों के सन्दर्भ में है कि क्या हमारे जीवन मूल्य बदल गए हैं? इस समय मृत्यु भय इस कदर हावी है कि हर तरफ जीवन को बचा लेने की कवायद जारी है। यह एक ऐसा पाठ है जो विश्व भर ने लिया होगा। कुछ ही दिनों में हमारी प्राथमिकताओं की सूची में वृहद रूप से बदलाव आया है और हमने यह भी सीखा कि जीवन जीना उतना भी कठिन नहीं जितना हम और आप माने बैठे थे! आखिर हमारी मूलभूत आवश्यकतायें हैं ही क्या? रोटी, कपड़ा और मकान के सिवा! कोरोना महामारी के इस घातक समय में जिसे देखिये वह बस एकमुश्त की रोटी और एक सुरक्षित छत से अधिक कुछ नहीं चाहता! हम कम खर्च में जीना भी सीख गए हैं।

हमने यह भी सीखा कि सुख की परिभाषा क्या है और क्या है जीवन की जरूरतें? वो जिन्होंने भौतिक सुख सुविधाओं और भोग विलास को ही सच्चा सुख मान लिया था उनके लिए सुख के मायने बदल गए हैं। अब वे सुरक्षित और स्वस्थ रहने से अधिक आवश्यक किसी बात को नहीं मानते वरना सोचिये कि दिन-रात समाज के बीच रहने वाले लोग बाहर निकलने से स्वयं को यूँ कभी नियंत्रित कर पाते! अब ग्रॉसरी खरीदते समय चटकारे से कहीं ज्यादा दाल-रोटी की फ़िक्र सताती है। 'दाल-रोटी खाओ, प्रभु के गुण गाओ' को हम सब इस समय जी रहे हैं। असल में जीवन तो सदा से ही सरल रहा है, बाजारवाद और अनौचित्यपूर्ण प्रतिस्पर्धा ने इसे जटिल बना दिया। काम और बातें ऑनलाइन हो रही हैं। आप कह सकते हैं कि मनुष्य की परस्पर दूरी बढ़ रही है लेकिन यह भी समझिये कि कुछ रिश्ते परवान चढ़ नई ऊंचाइयों को भी छू रहे हैं। कहीं दूरियां प्रेम को घटा देती हैं तो कहीं इसे स्नेह के कोमल धागे से बाँध आसमान की छत पर चुपके से टांक भी आती हैं।

क्या हममें आत्मनिर्भरता आई? यह प्रश्न आर्थिक आत्मनिर्भरता से सम्बंधित नहीं बल्कि घरेलू कार्यों के बारे में है। इसका जवाब इस समय घर में अकेले रह रहे पुरुषों और गृहिणियों से बेहतर कोई नहीं दे सकता! कितने ही पुरुष और युवा इस समय कहीं किसी दूसरे शहर में हैं। किसी भी तरह के हेल्पर्स की आवाजाही इन दिनों बाधित है तो ऐसे में क्या करें वो? झक मारकर ही सही पर जैसे-तैसे सीख रहे हैं। यह अभी के लिए बहुत कठिन समय है लेकिन उनका भविष्य आगे थोड़ा सरल हो जाएगा। गृहिणियाँ काम से थककर परेशान हैं लेकिन टाइम मैनेजमेंट सीखती जा रहीं हैं, इसका सबसे प्रमाण और क्या होगा कि हेल्पर्स के बिना भी उनकी मुस्कान सलामत है।

अब अंतिम बात कि क्या प्रकृति का यह प्रकोप और इससे सबक लेना जरुरी था? इसका उत्तर 'हाँ' होना नहीं चाहिए था पर दुर्भाग्य से है। मनुष्य ने जिस तरह और जितनी क्रूरता से प्रकृति का दोहन किया है, उसके प्रतिशोध में यह प्रकृति की Self Healing Process कही जा सकती है। इसके लिए हमें जो कीमत चुकानी पड़ रही है वह अत्यंत ही दुखद है। न जाने ये कौन सा नियम है कि सजा निर्दोष ही पाता है! कितनों के परिवार खत्म हो गए, घर उजड़ गए, नौकरियां छूट गईं और तबाही का ये मंजर अब तक थमा नहीं है। हमारी सावधानी, स्वच्छता और आत्म संयम के साथ जीना सीखने की कला ही अब इससे जीत सकती है लेकिन बदले में हमें भी प्रकृति को यह वादा देना होगा कि इसे विनाश और प्रदूषण से बचाना अब हम सबकी नैतिक जिम्मेदारी है। गर हम ये

न कर सके तो अगली बार चेतावनी दिए बिना ही ये प्रकृति हमें एक झटके में पलट देगी। बेहतर है कि हम प्रकृति से प्रेम करें, इसका और अपना दोनों का ख़्याल रखें।

कमी समय की है या प्राथमिकता की?

बहुत दिनों बाद उसे फोन लगाया था। बीते दिनों में न जाने कितने ही संदेश आते रहे और मैं 'हाँ, आती हूँ' कहती रही। जब भी पूछा कोई खास बात है क्या? तो उसका यही जवाब, 'नहीं, बस ऐसे ही बात करनी थी' सुनकर मैं निश्चिंत हो जाती। मैं चाहती थी जब भी उससे मिलूँ तो पूरा समय दे सकूँ, मेरी अपनी परेशानियाँ/व्यस्तताओं से निकल एकदम तसल्ली के साथ। पर जीवन रहते ये तसल्ली, ये आराम मिलता ही कितने पलों का है! यूँ कई मुलाकातों के कारण मुझे, उससे एक अपनापन-सा अवश्य हो गया था। मुझे देखते ही उसकी आँखें खिल उठतीं और दौड़कर कमरे के दरवाजे तक आ जाती। जाना मुझे ही होता हर बार, उसके लिए संभव नहीं था। ये आश्रम के कानून का उल्लंघन तो नहीं था, पर वो अकेली कभी कहीं गई ही न थी। उसकी उम्र 67 वर्ष थी, छद्म नाम देने से बेहतर है कि नाम ही न दूँ। सोचो तो बुरा लगता है कि मैं उसे आंटी जी क्यों नहीं कहती थी, 'उन्हें' पसंद नहीं था। शायद अपने जीवन में एक दोस्त की तलाश थी उन्हें, जिसे आंशिक भर सकी थी मैं। उन्होंने कसम दी थी मुझे, कभी भी आंटी/दीदी वगैरह न बोलने की। यूँ मैं इन बातों पर रत्ती भर भी विश्वास नहीं करती लेकिन यदि कोई दिल से इन्हें मानता हो तो उसका मान अवश्य रखती हूँ। अतः मैं उसे मोटूराम, चंपा, चमेली कुछ भी कहकर पुकारती थी और वो बेसाख़्ता खिलखिला उठती थी।

इतने महीनों बाद उसे कॉल किया था, जाने अभी कहाँ होगी! ऐसे में फोन पर किसी पुरुष की आवाज सुनकर एकबारगी तो मन झट से कॉल कट करने का हुआ, फिर भी हिम्मत जुटाकर अपना परिचय दिया। उधर से सपाट-सा जवाब मिला 'अरे, मम्मी ने आत्महत्या की कोशिश की थी। अभी हॉस्पिटल में हैं, कोमा में। भगवान जाने बचेंगी भी कि नहीं! मैं तो पैसा बहाते थक चुका हूँ।' न केवल इस बात से, बल्कि जिस लापरवाही और झल्लाहट के साथ ये कही गई थी; उसने मेरा दिमाग ही खराब कर दिया।

यूँ इस तरह की घटनाओं का ये पहला अवसर नहीं था, अपने उत्तरदायित्व को पूर्णरूपेण निभाने के बाद कुछ माता-पिता के हिस्से में यही बेरुखी आते मैं कई बार देख चुकी थी। इसीलिए दुःख इस बात का अधिक रहा कि उसने हॉस्पिटल के नाम के अतिरिक्त और कोई भी विवरण देने से इंकार कर दिया। संभवतः ग्लानिवश अपना नाम छुपाना चाहता होगा। खैर! डॉक्टर से बात करने पर पता चला था कि वे अब एक-दो दिन की ही मेहमान हैं। ठीक यही हुआ भी! तीसरे ही दिन वे चल बसीं।

मैंने उनकी आँखों में जीवन के प्रति एक ललक देखी थी, उनकी मुस्कान के पीछे छिपा दर्द भी बखूबी समझती थी। यूँ तो वहाँ वृद्धाश्रम में रहने वाले सदस्यों में से एक प्रतिशत ही ऐसे थे जो कि स्वेच्छा से आये थे। शेष सभी की अपनी अलग कहानी थी, हालाँकि थीम और क्लाइमेक्स तो सबका एक ही था। लेकिन तमाम कठिनाइयों के बावजूद भी मेरी इस दोस्त को जीवन जीना और खूब हँसना-हँसाना भी आता था। 'आत्महत्या' को वो भी बेहद मूर्खतापूर्ण कृत्य माना करतीं थीं। न जाने क्या हुआ होगा! यह प्रश्न मुझे अब तक खाये जाता है। सबको दिलासा देने की मेरी कवायद और कुछ अधिक न कर पाने की मेरी विवशता ने इस बार मुझे वाकई तोड़कर रख दिया था। जिससे उबरने में मुझे महीनों लगे।

वृद्धजन के साथ घटित यह कोई एकमात्र किस्सा नहीं है। ऐसी अनगिनत घटनाएँ हमारे समाज में प्रतिदिन घटा करती हैं। पर प्रायः इन्हें सरोकार की श्रेणी में रखने से लोग बचते रहे हैं। वैसे भी स्वार्थ, राजनीति और सोशल मीडिया की लिप्तता ने संवेदनाओं को एक क्लिक भर से व्यक्त कर इतिश्री पा ली है।

आपकी स्मृति में हो शायद कि आशा साहनी नाम की एक वृद्ध महिला का शव, उन्हीं के घर से उनके बेटे को एक वर्ष बाद मिला था, जब वह अपनी माँ से मिलने आया था। यह कोई सनसनीखेज समाचार नहीं बल्कि पूरे समाज के गाल पर एक झन्नाटेदार तमाचा था। आखिर हमारी संवेदनशीलता को हुआ क्या है? समाज से इतनी विरक्ति? इतनी आत्म-केन्द्रीयता कि पड़ोसी की मृत्यु की भनक तक नहीं? घर अंदर से बंद हो और महीनों न खुले, तो किसी को चिंता न हुई? किसी का मन विचलित न हुआ? क्या हम इतने पाषाण हो गए हैं कि किसी की अनुपस्थिति हमें चिंतित नहीं करती?

कहते हैं कि करोड़ों का अपार्टमेंट था। उसमें रहने वाली वृद्ध महिला की भूख से मौत? ऐसी मृत्यु अचानक तो नहीं होती! बेटा अमेरिका में था पर क्या पूरे भारत में उनका एक भी रिश्तेदार या मित्र नहीं था जो कभी हालचाल पूछता?

जब हम अपने कर्त्तव्य ही नहीं निभा पाते तो किस मुँह से अधिकारों की बात करते हैं?

बड़े-बड़े उद्योगपति घरानों की पारिवारिक कलह भी हमेशा से सुर्खियाँ बटोरती आई है पर कितना अविश्वसनीय और दुःखद है कि जिसने दिन-रात एक कर अपने सपनों का महल साकार किया, आज वही पाई-पाई को विवश है! क्या यही दिन देखने के लिए माता-पिता अपने बच्चों की परवरिश करते हैं!

ये कटु सत्य हम सबको निगलना ही होगा कि वृद्ध हर जगह

उपेक्षित हैं, घर के भीतर भी और बाहर भी! खासतौर से तब जब वो अपनी शारीरिक विषमताओं से संघर्षरत हैं।

क्या आपको नहीं लगता कि जीवन की भागदौड़ में सभी आवश्यक कार्य छूटते जाते हैं? कभी बच्चों को उतना समय नहीं मिल पाता जिसके वो हकदार हैं तो कहीं वृद्ध माता-पिता स्वयं को उपेक्षित महसूस करते हैं! आखिर यह समय जाता कहाँ है? किसके लिए जी रहे हैं हम सब? क्या यह मात्र रोटी की दौड़ है या हम कुछ और सिद्ध करना चाहते हैं? कुछ ऐसा जिसे हमारे जाने के बाद कोई न याद रख सकेगा! अंत समय में हम गिनेंगे कि सबसे आगे बढ़ने की दौड़ में कितना कुछ पीछे छोड़ आये हैं कितने अधूरे काम, अधूरे पल, अधूरा साथ, अधूरी मित्रता और भी न जाने क्या-क्या टूटा, क्षत-विक्षत, अपूर्ण! पर तब बीते समय में लौट पाना संभव न होगा! लेकिन एक प्रश्न जरूर कौंधेगा, शायद कचोटता भी रहे कि यह समय की कमी थी या प्राथमिकता की?

समय लौटकर नहीं आता!

समय कब और कैसे बीतता जाता है, पता ही नहीं चलता। प्रतिदिन हम कुछ कार्यों को यह कहकर टाल देते हैं कि अभी समय नहीं, पर क्या यही एकमात्र सत्य होता है? वास्तविकता तो यह है कि वे कार्य या तो हमारी प्राथमिकता सूची का हिस्सा नहीं या फिर उनको करने की हमारी मंशा ही नहीं होती। तो फिर इस सच्चाई को स्वीकार कर दूसरे पक्ष को ये बात निस्संकोच कह ही देनी चाहिए। इससे आप स्वयं को दोषी मानकर हुए अपराध-बोध और किसी की नजरों में गिरने से भी से बच सकते हैं। छोटी-छोटी बातें ही मन में घर कर लेती हैं, कब यह गाँठ बनकर रिश्तों को फाँसी दे दे, कोई नहीं जानता! सच की स्वीकारोक्ति से बेहतर कुछ भी नहीं,इससे मात्र अपना ही नहीं, कइयों का जीवन संवर जाता है और संबंधों की मधुरता भी यथावत रहती है।

खून का रिश्ता, स्वतः ही मिलता है, इसे चुनने की सुविधा नहीं होती, पर आश्चर्यजनक बात है, कि इसमें दुविधा भी नहीं होती। जो है, जैसा है...स्वीकार करना ही होता है। कर ही लिया जाता है। परिवार इसी को तो कहते हैं, जहाँ अलग-अलग सोच, पसंद और प्राथमिकताएँ होते हुए भी सब साथ रहते हैं। यूँ खूब लड़ते-झगड़ते भी हैं पर मौका आने पर हर बाधा से टकराने को एक हो जाते हैं। कितने भाग्यवान हैं वो लोग, जो अपने परिजनों के साथ हर सुख-दुख साझा करते हैं, परस्पर प्रेम से रहते हैं, एक-दूसरे को समझते और मदद करते हैं, उनके लिए पर्याप्त समय भी देते हैं। वहीं कुछ घरों में सब कुछ होते हुए भी समय का अभाव रहता है, सब अपने-आप में ही व्यस्त। कुछ व्यस्त पहले से ही थे और कुछ ने स्वयं को व्यस्तता की रस्सी से भीतर तक बाँध रखा है। ये परिवार नहीं, एक 'सामाजिक व्यवस्था' के तहत निभाए जाने वाले रिश्ते बनकर ही प्रारंभ होते है और उसी अवस्था में प्रायः दम भी तोड़ देते हैं!

किसी की खुशी उसका चेहरा ही बयाँ कर देता है जहाँ मन की प्रसन्नता शब्दों की ध्वनियों पर थिरकती नजर आती है, तो कहीं होठों पर खनकती हँसी, आँखों की नमी को दम साधे, भीतर ही खींचे रखती है। हृदय से निकलती कितनी ही चीखें अंदर सांय-सांय करती अपनी ही आवाजों के शोरगुल में दबकर रह जाती हैं और फिर दर्द बाहर निकलने का रास्ता तलाशने लगता है। यही जीवन है। इसमें बचपन की बातें हैं, स्कूल-कॉलेज के किस्से हैं, कितने खुशनुमा पल हैं, कभी कोई कमी अखरती है तो कभी बहुतायत से मन ऊबने लगता है। कितनी सुंदर यादें हैं, कहीं कोई भूला हुआ, कभी कोई

याद आता है। गीतों की सुमधुर बारिश में हर कोई गुनगुनाता है। जीवन है, तो अहसास है। अहसास है तो प्रेम है। प्रेम में सुखद अनुभूति है और भीषण पीड़ा भी। कभी साथ रहने तो कभी एकांत की चाहत में दिल बेचैन हो उठता है, परन्तु ये दोनों ही स्थितियाँ, चाहने भर से नहीं मिलतीं! मन की बात किससे कही जाए! ऐसे में देखें तो रचनाकार, कितना भाग्यवान है कि उसे अपने भावों को पूरी बेबाकी और ईमानदारी से लिख पाने की स्वतंत्रता है। लिखना तमाम पीड़ाओं से मुक्त हो जाना भी है।

उम्मीद को लिखो, खुशी को लिखो, उदासी को लिखो,अभाव को लिखो, जलधार को लिखो, वक्त की गहरी मार को लिखो। प्रेम को लिखो, विरह को लिखो, दिन के चारों पहर को लिखो, दुनिया में घटते कहर को लिखो। समाज के खोखलेपन पर लिखो, झूठी मर्यादाओं पर लिखो, बढ़ते अहंकार पर लिखो, कुंठित, टूटते लोगों पर लिखो, गरीब के झोंपड़े पर लिखो। तुम लिखना जरूर क्योंकि तभी कोई समझ सकेगा। तुम कहना जरूर क्योंकि तभी कोई सुन सकेगा। रोना जरूर, कि दुःख बह जाता है। हँसना भी है कि इससे मन खिलखिलाता है।

समय की आपाधापी में बहुत कुछ छूटता चला जाता है। हमारी कोशिश होनी चाहिए कि भीतर बचपन जीवित रहे, अपने-आप से भी एक रिश्ता बना रहे। परिवार रहे, दोस्ती रहे, संबंधों में मधुरता रहे, स्वस्थ जीवन रहे! न रहे कहीं तो...बस कोई अफसोस! ध्यान से देखें, आत्मविश्लेषण करें, पुनर्विचार करें कि 'समय' की कमी, कहीं भावनाओं का अवरोहण तो नहीं? क्योंकि जब तक हम जीवित हैं समय भी तभी तक हमारे पास है! हमारे मिट्टी में मिलते ही, समय भी रीत जाएगा! उसके बाद, कौन देख सका है! कौन मिल सका है! लाख चाहने पर भी, उम्र भर पुकारने पर भी न तो बीता हुआ समय लौटकर आता है, और न ही इंसान!

जीवन से बेहतर कोई विकल्प नहीं!

हम सुख को पल भर में भूल, दुःखों को उम्र भर ओढ़ लेने वाली पीढ़ी का प्रतिनिधित्व करने लगे हैं। पीड़ा का मनन, 'पीड़ा' से भी कहीं अधिक दुखदायी होता है लेकिन यह जानते-समझते हुए भी हम कहाँ इससे बच पाते हैं! दरअसल यह हमारी आदत में आ चुका है और इसीलिए शायद ही कोई ऐसा हो जिसने कभी मृत्यु की कल्पना न की हो पर आश्चर्य यह भी कि जीने का मोह छूटता नहीं कभी! कमजोर पल जितना अवसाद के करीब ले जाते हैं, वहीं जिंदगी भी किसी जिद्दी बच्चे की तरह कसकर हाथ थामे साथ चलती है।

इधर आत्महत्या के कई किस्सों ने हृदय विचलित कर दिया है। कभी-कभी लगता है कि 'आत्महत्या' एक प्रकार की हत्या ही है। अंतर बस इतना ही है, कि इसमें इंसान अपने मरने के तरीके स्वयं चुनता है माहौल तो दूसरे या उसके अपने ही तैयार कर देते हैं। 'हत्या' में, हत्यारे को उस व्यक्ति की उपस्थिति सहन नहीं होती या घृणा अपनी सारी सीमाएँ पार कर लेती है इसलिए वह उसे मार डालता है। 'आत्महत्या' में, इंसान को लगने लगता है कि उसकी उपस्थिति के कोई मायने ही नहीं! या फिर प्रतिकूल परिस्थितियों के सामने वह असहाय एवं लाचार अनुभव करने लगता है। प्रथम दृष्टि में ही वह कमजोर और मूर्ख प्रतीत होता है।

स्वभावतः मैं हमेशा से ही 'आत्महत्या' करने वालों से नफरत करने वाली और इस कृत्य की घोर विरोधी रही हूँ क्योंकि मुझे लगता आया है कि उन्हें अपनों की रत्ती भर परवाह नहीं होती इसीलिए ये कायरतापूर्ण कृत्य कर बैठते हैं। लेकिन कभी-कभी एक अप्रत्याशित-सी सहानुभूति भी होती है उनसे कि हो सकता है, उन्हें भी यही महसूस होता हो कि अपनों को ही उनकी चिंता नहीं! अगर मृत्यु से पहले, कोई उन्हें यह अनुभूति करा देता कि उनकी उपस्थिति कितनी महत्त्वपूर्ण है तो वे संभवतः कभी ऐसा करने का सोचते तक नहीं!

ऐसे में दिल से यही आवाज आती है कि काश! कोई किसी को कभी ऐसा करने को विवश न करे! दिल से न सही, एक 'भ्रम' की तरह भी कोई, किसी का हमेशा के लिए हो सके तो कुछ जिंदगियाँ, कुछ वर्ष और जी सकती हैं! अकेलापन भी मार ही देता है, इंसान को! ये मौत भीतर ही होती है, हर रोज! कभी समाज का भय, तो कभी व्यक्ति-विशेष से बिछोह का। कभी किसी की कसौटी पर खरा नहीं उतर पाना बुरा लगता है तो कभी अपने ही उत्तरदायित्वों के निर्वाह में कमी सी लगती है। 'अपेक्षा' और 'उपेक्षा' दोनों ही हर रूप में घातक हैं। इनसे बच गये, तो भाग्यशाली हैं।

'एकला चलो रे' सुनने में अच्छा जरूर लगता है, पर कौन, कब तक, कितनी दूर तक अकेला चल पाया है, इसका हिसाब-किताब कहीं नहीं! हाँ, जिसने भी इस मंत्र को समझ लिया, वही जी गया! कतरा-कतरा जिंदगी तो रोज ही बिखरती है और रोज ही उसे समेटना भी जरूरी है। बहुत रहस्यवादी है ये दुनिया और इसमें बसे सभी लोग। किस्मत वाले हैं, वो जो इसमें खुलकर हँस लेते हैं! पर कौन जाने, उस हँसी की तहों में, रोज कितने दर्द दफन होते हैं!

क्या 'आत्महत्या' जीवन की मुश्किलों से भाग जाने का एक आसान उपाय है या कि आखिरी? मैं अब कारणों पर विचार करने लगी हूँ। कुछ वर्ष पूर्व दसवीं कक्षा के कुछ विद्यार्थियों की इसी प्रवृत्ति को दूर करने के लिए मैं उनके साथ कुछ माह तक जुड़ी रही थी। यह घटना अब उम्र तक नहीं देखती। बच्चों में यह सोच दूर करना उतना मुश्किल नहीं था क्योंकि अधिकांशतः यह माता-पिता का दबाव या भविष्य के प्रति अतिरेक चिंता का दुष्परिणाम होता है। जिसे बातचीत से समझाया जा सकता है। परेशानी उन बड़ों की अधिक है जो कि समझते हुए भी नहीं समझ पाते।

यह तो स्पष्ट ही है कि घोर निराशा और अवसाद के दौर से गुजरते हुए ही इंसान आत्महत्या कर बैठता है और यह उन एक-दो पलों के निर्णय का तुरत परिणाम ही होता है, जहाँ सोचने-समझने की गुंजाइश खत्म हो चुकी होती है। हर बात को समाज, दोस्तों और परिवार पर थोपना बेहद आसान है और इन्हें दोषमुक्त नहीं किया जा सकता पर आजकल स्वार्थ, लोलुपता, ईर्ष्या और अपनी-अपनी दुनिया में रमे रहने का जो दौर चल पड़ा है वहाँ किसी से भी आवश्यकता से अधिक अपेक्षा रखना, उपेक्षा को जन्म देता ही है। हम इसी सोच में रोते रहते हैं कि हमने किसके लिए, कब, क्या और कितना किया! किसके लिए अपनी इज़्ज़त को भी ताक पर रख पूरे जमाने से लड़े! किरकिरी हुई वो अलग और बदले में मान तक न मिला। उस वक़्त हम यह भूल जाते है कि कुछ रिश्ते मतलब के लिए ही बनते हैं, कुछ सुविधानुसार। जैसे ही आप उनकी राह में अड़चन बने, आपने उनकी हाँ में हाँ मिलाना बंद किया या फिर अब उनकी आपसे जुड़ी कोई भी अपेक्षा न रही या सभी पूरी हो चुकी हैं, वैसे ही आप उनकी प्राथमिकता सूची से बाहर फेंक दिए जाते हैं।

ये बदलते समय, बदलती मानसिकता, बदलती सोच, बदलते विश्वास, बदलती आस्थाओं और सेल्फी का दौर है। अति संवेदनशील और भावुक लोग अब मूर्ख की श्रेणी में गिने जाने लगे हैं। उनके आँसू स्मार्ट जेनरेशन को चिढ़ देते हैं क्योंकि उनके पास न उन्हें

पोंछने का वक़्त है, न हृदय को टटोलने का। एक दौड़ में सभी भागे जा रहे हैं, कब और कहाँ ठहरेंगे इन्हें भी नहीं पता। पर इतना तय है कि समय बीत जाने के बाद ही सही पर ठहरेंगे अवश्य! लेकिन संवेदनशील लोगों को इन्हें भूलना और क्षमा कर आगे बढ़ना सीखना ही होगा। यदि किसी इंसान की जिन्दगी में आपकी कोई अहमियत नहीं तो आप भी जितना शीघ्र हो सके उसे अपने मन-मस्तिष्क से शीघ्र बाहर निकाल फेंकिए। वो तो मजे में कहीं घूम रहे होंगे और इधर आप उनकी याद में औंधे पड़े हैं।

सोशल मीडिया पर यही स्थिति 'प्रेम' की है जो अब एक साप्ताहिक बीमारी की तरह हो गया है। इसमें एक इंसान कई लोगों को एक साथ प्रेमजाल में फाँसता है और फिर आराम से अपने अगले 'शिकार' की तरफ बढ़ जाता है। ऐसे मूडी और टाइम पास करने वाले प्रेमी/प्रेमिकाओं के लिए अपनी जान गँवाना दुःखद नहीं हास्यास्पद है। जितना जल्दी हो सके, इनके चंगुल से बाहर आइये। ऐसे धोखेबाजों के लिए अपनी जान क्यों देना, जिन्हें जीते-जी ही आपकी परवाह नहीं

होता यह है कि प्रेम में अंधा इंसान हर बात में हामी भर देता है, दूसरे पक्ष को इसकी लत हो जाती है और बाद में एक प्रतिरोध भी उसको बौखला देता है क्योंकि उसने स्वयं को आपका मालिक समझ लिया होता है। अहंकारी इंसान रिश्ते निभाना नहीं जानते।

बेहतर हो कि किसी दम्भी इंसान से करीबी महसूस हो तो उसे शुरू में ही परख लें। बस एक बार उससे मदद मांगकर देखिये, अपना दुःख बाँटिये, अपनी सफलता पर उसकी प्रतिक्रिया देखिये, उसकी असलियत का पर्दाफाश वो खुद ही कर देगा। उसके बाद ही तय करें कि वो आपका कितना अपना है और किन परिस्थितियों में भाग खड़ा होता है। कौन अपना जीवन ऐसे ईर्ष्यालु और घटिया इंसान पर न्योछावर करना चाहेगा! काश, आत्महत्या करने वाले देख सकते कि जिसके लिए मौत चुनी वो एक नजर देखे जाने के योग्य भी नहीं थे।

ईश्वर की असीम अनुकम्पा से हवाओं की सुगंध अभी गई नहीं, प्रदूषण चाहे लाख बढ़ गया हो पर साँसें अनवरत चल सकतीं हैं, कुछ मासूम चेहरों को हमारे स्नेह की जरूरत अब भी है, हमें उनकी मुस्कान बनना है। कुछ भी न हो तो अपने-आप के लिए जीना है। बिना किसी चाहत, बिना किसी उम्मीद, बिना किसी अपेक्षा के! नए सिरे से जिन्दगी जीना शुरू करना ही होगा। यूँ भी आपकी मृत्यु से उन्हें कोई अंतर नहीं पड़ने वाला, जिनके लिए आप हमेशा ही एक अनावश्यक सामान भर बनकर रह गए। उन्हें अपनी जिन्दगी

से उनके सामने ही बाहर निकाल फेंक शान से जीना सीखना होगा और जिन्हें आपकी कद्र है उनको मान देना तो बनता ही है।

खैर... खुदा करे, आपके सभी अपने, आपके करीब और हमेशा सलामत रहें! जिंदगी, जीने के लिए ही तो बनी है! इससे कैसी नाराजगी! जीवन से बेहतर कोई विकल्प नहीं'!

पुरुष को भी होती है पीड़ा

ये तो हम जानते ही हैं कि 19 नवंबर को 'अंतरराष्ट्रीय पुरुष दिवस' (International men's day) के रूप में विश्वभर में मनाया जाता है। इस विशिष्ट दिवस की संकल्पना पुरुषों के साथ होने वाली असमानता, हिंसा, उत्पीड़न, और शोषण को रोकने के लिए तथा उन्हें समान सामाजिक अधिकार दिलाने के उद्देश्य से की गई थी। प्रत्येक वर्ष इसकी थीम अलग होती है।

चूंकि 'अंतरराष्ट्रीय महिला दिवस' कई वर्षों से चला आ रहा था अतः लैंगिक समानता को बढ़ावा देने के लिए इस दिवस की मांग उठी थी। अवसाद के मामलों में वृद्धि देखते हुए भारत में 2007 में पहला 'मानसिक स्वास्थ्य कानून' बना तथा इस दिवस को मनाना भी 2007 से ही प्रारम्भ हुआ।

यूनेस्को द्वारा सहयोग प्राप्त इस दिवस को मनाने के मूल में यह है कि घर, परिवार और समाज में उनके सकारात्मक सहयोग को सराहा जाए। उनके भावनात्मक पक्ष पर ध्यान दिया जाए तथा उनकी स्वास्थ्य संबंधी एवं मानसिक परेशानियों पर भी चर्चा हो।

स्पष्ट है कि हम हर कार्य की अपेक्षा सरकार से नहीं कर सकते हैं और जब यह पारिवारिक स्तर पर प्रारम्भ हो तभी एक स्वस्थ, संतुलित और ऊर्जावान समाज की सोच सार्थक होगी।

हम अपना सहयोग निम्न प्रकार से दे सकते हैं–

मांओं को चाहिए कि वे अपने बेटे को रोते हुए देख कभी ये न कहें कि 'अरे! लड़का होकर रोता है!' रोने दीजिये उसे। ऐसे ही भावनाएं निकलती हैं और दुःख बह जाता है। हमें उन्हें पाषाण एवं संवेदनहीन नहीं बनाना है कि वे किसी का दर्द समझ ही न सकें कभी।

घर के कार्यों में भी पूरा सहयोग लें। जिससे बाद में बाहर जाएं तो सब मैनेज कर सकें।

उन्हें स्वस्थ रहने की सलाह दें, खलीनुमा बनने की नहीं मजबूत दिखने के चक्कर में कई बार लड़के घातक दवाइयों एवं स्टीरॉइड के फेर में पड़ जाते हैं। ऐसे में जिम वाले इसका पूरा लाभ उठाने में नहीं चूकते।

स्त्रियों को चाहिए कि वे पुरुष को अपनी हर समस्या का 'समाधान केंद्र' न समझें। कभी उनकी भी सुनें। परेशानियां उनके जीवन में भी उतनी ही हैं। आपकी समस्याओं और रोज के बुलेटिन में उलझ वे अपना गम बांट ही नहीं पाते कभी। दर्द उन्हें भी होता है, डर उन्हें भी लगता है पर आपकी रोज की हाहाकार ने उन्हें सुपरमैन, बैटमैन और चाचा चौधरी बना डाला है।

केवल स्त्रियां ही नहीं, पुरुष भी शारीरिक-मानसिक शोषण, घरेलू हिंसा, उत्पीड़न एवं भेदभाव के शिकार होते हैं। परन्तु हमारे समाज का ढांचा कुछ ऐसा है कि इनके दर्द भरे किस्से दबा दिए जाते हैं या फिर उनका उपहास उड़ाया जाता है। सुना होगा न– 'अरे, कैसा मरद है रे तू? एक लड़की से थप्पड़ खा लिया!', 'कहां गई तेरी मर्दानगी!' अरे भई! वो भी मनुष्य ही है कोई बुलेटप्रूफ जैकेट नहीं! उसके मस्तिष्क में अपने अनर्गल विचारों का कचरा मत डालिये। पुराने डायलॉग भूलकर समझ लीजिये कि 'मर्द को दर्द भी होता है' अतः उनकी पीड़ा को समझ उस पर प्रेम का मरहम रखिये। तानों का तम्बूरा न बजाइये।

आत्महत्या और हत्या का अनुपात भी इन्हीं का सर्वाधिक है। आखिर यह अवसाद कहां से उपजा है? इनकी कुंठाओं की जड़ों में जाना आवश्यक है और उससे पहले इन्हें समझना जरुरी है। देखिये कहीं आपकी भौतिक अभिलाषाओं का बोझ इन पर भारी तो नहीं पड़ रहा।

आपके वेकेशन प्लान्स उनपर लादे नहीं उनकी छुट्टियों से सामंजस्य स्थापित करने का प्रयास करें। इधर आप घूमने को लेकर झगड़ रहीं हैं और उधर वो बॉस से परेशान है। हां, यदि वे कामचोर हों और हर माह दस छुट्टियां लेकर घर बैठें तो उन्हें देश के विकास का पाठ अवश्य पढ़ाइए।

यूं मैं स्त्रीवादी/पुरुषवादी सोच न रखकर मानववादी सोच की पक्षधर हूं क्योंकि एक अच्छे समाज के निर्माण के लिए स्त्री-पुरुष दोनों की ही उपस्थिति की महत्ता और सहयोग के बराबर मायने हैं। यद्यपि ऐसा समाज कल्पना भर है।

वर्तमान परिप्रेक्ष्य में साहित्यकारों का दायित्व

चित्रकार, मूर्तिकार, साहित्यकार या किसी भी कलात्मक कार्य से संबद्ध लोग मूलतः रचनाकार ही होते हैं और किसी भी रचनात्मक व्यक्ति का एकमात्र धर्म सृजन ही है। वह निर्माण में विश्वास रखता है, विध्वंस में नहीं! उसका जन्म नकारात्मकता के सारे भावों को दूर कर समाज में सकारात्मकता और आशावाद के प्रचार-प्रसार के लिए हुआ है। उसका सतत प्रयत्न होता है कि वह ईर्ष्या, द्वेष के विकारों को दूर करे। उसकी आँखों में सृष्टि की सुंदरता का बखान और हृदय में इसकी कुरूपता को विस्तार देने वाले तत्त्वों को नष्ट कर देने का भाव संचित होता है। उसकी भाषा सभ्य, सुसंस्कृत और सारगर्भित होने की माँग करती है कि लोग उसका अनुसरण करें। उसके लिखे शब्द, उसकी बनाई मूर्ति या किसी चित्र में भरे रंग प्रेरणास्पद हों और लाख निराशा के बाद भी उनमें जीवन के प्रति उल्लास झलकता हो तो ही इस क्षेत्र से जुड़े लोगों का जीवन सार्थक है क्योंकि इनका लक्ष्य सामाजिक बुराइयों को दूर करना, अव्यवस्थाओं को इंगित कर निराकरण के विभिन्न उपाय बताना है।

साहित्य, यथार्थ की कंटीली भूमि में उगी फसल है, जो हर तरह के मौसम से प्रभावित होती है। ये सामाजिक उथल-पुथल, परिवर्तित मूल्यों और बदलती संवेदनाओं की शाब्दिक अभिव्यक्ति है। रचनाकार के शब्दों की उष्णता, उसके संवेदनशील हृदय की प्रतिकूल परिस्थितियों से उपजी कड़वाहट का जीता-जागता हस्ताक्षर होती है और स्नेह, सुन्दर अनुभूतियों की शाश्वत प्रस्तुति है। मूलतः उसकी लेखनी पर सामाजिक परिप्रेक्ष्य का प्रत्यक्ष प्रभाव पड़ता ही है।

सोशल मीडिया के बढ़ते प्रचार-प्रसार के चलते वर्तमान परिप्रेक्ष्य में साहित्यकारों का दायित्व और भी विस्तृत हो चला है। अतः अपनी लेखनी को स्वर देते समय उनके लिए निम्नलिखित बिंदुओं का सतत स्मरण आवश्यक है-

घृणा के बीज को खाद-पानी न दें- राष्ट्रहित की बातें करने के नाम पर कई बार लेखक स्वयं ही विभिन्न धर्मों, जातियों के बीच वैमनस्यता बढ़ाने वाले निरर्थक तथ्य प्रस्तुत करने लगते हैं जिनका यथार्थ से कोई सम्बन्ध नहीं होता! यह देश सबका है और यदि कहीं कट्टरता पर जोर देना है तो वह 'कट्टर भारतीय' बनने पर दिया जाना चाहिए। इससे कम या अधिक कुछ नहीं!

सत्य को निर्भीक होकर लिखना- प्रत्येक समय काल में साहित्यकारों का सर्वप्रथम दायित्व निर्भय, निडर होकर सत्य के पक्ष में खड़ा होना है। उसे समाज के सामने परिस्थितियों को जस का तस सटीक रूप में रखना है। वह किसी व्यक्ति विशेष, दल, सत्ता,

पक्ष या विपक्ष के हितध्अहित के लिए नहीं लिख सकता। प्रबल लेखनी के लिए नितांत महत्वपूर्ण है कि बिना किसी झुकाव या दबाव के वह केवल सच लिखे और उसी के समर्थन में ही सदैव अडिग खड़ी नजर आये।

अफवाहों के प्रचार-प्रसार से बचें– डिजिटल दुनिया के विकास का प्रभाव मीडिया और समाचार पत्रों पर भी पड़ा है। अब सूचना तत्काल दौड़ती हुई लोगों के दिमाग तक पहुँचती है। लेकिन इसकी अपनी चुनौतियाँ हैं। अफवाहें खबर बनकर घूमती हैं और उनको रोकना मुश्किल होता है। इससे डिजिटल मीडिया की विश्वसनीयता प्रभावित होती है लेकिन इसके बावजूद इसका विस्तार कोई नहीं रोक सकता है। पत्रकारों को इस क्षेत्र में आगे आकर काम शुरू करना चाहिए। डिजिटल मीडिया का प्रभाव तात्कालिक और बहुत गहरा ही नहीं बल्कि विश्वव्यापी भी होता है। चुनाव हो या व्यापार,यह हर निर्णय को प्रभावित करता है। ऐसे में वे साहित्यकार जो सोशल मीडिया पर पत्रकारिता भी करने लगे हैं उन्हें व्यक्तिगत महत्वाकांक्षाओं से परे होकर किसी भी गलत बात का खंडन एवं खुलकर विरोध करना चाहिए।

देश की स्त्रियों और बच्चों पर उसकी कलम अवश्य चले– किसी भी देश की समृद्धि और समुचित वातावरण को वहाँ की महिलाओं और बच्चों की स्थिति से समझ लेना चाहिए। यह लेखकों का दायित्व है कि समाज में रहने वाले प्रत्येक वर्ग की स्त्रियों और बच्चों की दैनिक परेशानियों पर लिखे और समस्या के समाधान के लिए अधिकारियों की नाक में दम कर दें। उसे प्रजा की आवाज, राजा तक हर हाल में पहुँचानी चाहिए।

गुटबाजी से बचे और अपनी राह स्वयं बनाए– सोशल वेबसाइट्स पर ऐसे सैकड़ों लोग मिल जाएँगे, जिनके यहाँ होने का एकमात्र उद्देश्य दूसरे को नीचा दिखाकर स्वयं आगे बढ़ना है। कुछ ने परस्पर गाली-गलौज और अभद्र भाषा का प्रयोग कर अपने-अपने मठाधीशों की 'छावनी' में स्थान पा लिया है। कुछ आलोचक बन गए हैं और कुछ बड़ी-बड़ी पत्रिकाओं में छपकर चर्चित भी हो चुके हैं। क्या ये मनन करेंगे कि इस लक्ष्य तक पहुँचने की सम्पूर्ण प्रक्रिया में इन्होंने लेखक प्रजाति की कितनी छीछालेदर की है? उनके प्रति उपजे सम्मान-भाव को कितना नीचे गिरा दिया है? इस सब में अगर किसी का नुकसान है तो वो सच्चे और ईमानदार लेखकों का है, जिनका गुटबाजी से कोई लेना-देना नहीं, जिन्हें चाटुकारिता आती नहीं और जो सत्य को सत्य की ही तरह पेश करते हैं।

एक दुखद पहलू यह भी है कि कुछ लेखकों की स्वार्थ के आगे घिग्घी बंध जाती है! गुटबाजी और एक-दूसरे को प्रमोट करने

की चाहत इनकी जुबां पर चुप्पी का मजबूत शटर गिरा देती है।

भाषा सारगर्भित एवं मर्यादित हो– साहित्य की भाषा, विषय और चिंतन में 'बोल्डनेस' जगह बना रही है। कहते हैं, 'हम वही खाते हैं, जो परोसा जाता है।' हिन्दी सिनेमा में द्विअर्थी संवाद या गानों के लिए यही तर्क दिया जाता रहा है। दर्शक फिल्मकारों पर सामाजिक पतन का दोष मढ़ते हैं और फिल्मकार इसे अपनी मजबूरी और 'यही चलता है', कहकर टाल देने का प्रयत्न करते हैं। ठीक यही स्थिति हिन्दी साहित्य की भी होती जा रही है। कुछ भी, कैसी भी, ऊल-जलूल भाषा में लिख देना और फिर उसका छप जाना बेहद प्रचलित हो गया है। भाषा के स्तर से अधिक ध्यान मार्केटिंग पर दिया जाता है। जितना अच्छा प्रचार, उसी के अनुपात में बिक्री तय होती है। साहित्यकार अब व्यापारी होता जा रहा है। प्रकाशक और उसके बीच गठबंधन-सा होने लगा है। भाषा, मुखपृष्ठ लील हो या अश्लील, इससे किसी को इतना अंतर नहीं पड़ता, बिकना प्राथमिकता है। 'साहित्यकार' बनना अब दो मिनिट मैगी जैसा है।

सामाजिक एवं चारित्रिक विकास में योगदान दे– साहित्य का हमारे चरित्र और सोच पर बहुत गहरा प्रभाव पड़ता है। हमारी इच्छा शक्ति, दृढ़ता, देश के लिए मर-मिटने की भावना और आत्म निर्माण में साहित्य गहरी भूमिका निभाता है। अच्छे विचार, अच्छा लेखन ऊर्जा-संचारक और प्रेरणा-स्रोत के रूप में कार्य करते है। कहा ही गया है, 'पुस्तक सच्ची मित्र होती है।' ऐसे में यह और भी आवश्यक है, कि लिखी जाने वाली भाषा पठनीय व स्तरीय हो। अश्लील साहित्य से व्यक्ति और समाज का उत्थान संभव नहीं, हाँ ये पतन का कारण अवश्य बन सकता है।

साहित्यिक चोरी का खुलकर विरोध करें– इस समय हम चोरी, उठापटक, गुटबाजी, धोखे, झूठ और कॉपी पेस्ट के उस दौर में लिखे जा रहे हैं जहाँ लेखक सर्वाधिक असहाय प्राणी नजर आता है। उसकी लिखी रचना मंच पर पढ़ी जाती है, पत्र-पत्रिकाओं में प्रकाशित करवा ली जाती है, तोड़-मरोड़कर उन्हें मौलिक रूप देने का प्रयास किया जाता है। बीते कई वर्षों से यह प्रचलन इतना अधिक देखने को मिल रहा है कि देश की अन्य समस्याओं की तरह हमारे सिस्टम ने अब इसे भी डाइजेस्ट करना सीख लिया है। लेकिन अब लगने लगा है कि तकनीक के इस युग में एक ऐसा भी समय आएगा जब मूल रचनाकार को ही अपनी रचना की मौलिकता सिद्ध करने के लिए दर-दर भटकना होगा और उसे अपनी कह देने वाला वाहवाही के प्रतिभाशाली दरबार में सम्मानित होगा।

ऐसी कई घटनाओं के हम समय-समय पर साक्षी होते भी रहे हैं कि यदि साहित्यिक चोरी का आक्षेप किसी नामी रचनाकार

पर लगा हो तो उसके प्रकाशक, मित्र बिना प्रमाण को देखे-समझे उसके साथ खड़े हो जाते हैं और पीड़ित पक्ष विवश, ठगा सा खड़ा रह जाता है!

विचारणीय बात यह है कि एक लेखक के हिस्से यूँ भी आता ही क्या है? इसके बाद भी वह मात्र आत्मसंतुष्टि हेतु सतत सृजनशील रहता है जिसे हम और आप 'स्वान्तः सुखाय' कहकर तसल्ली पा लिया करते हैं। अब वो लिखा भी कोई और उठा अपना लेबल लगा ले तो रचनाकार के लिए क्या शेष रहा?

आत्सम्मान को बनाए रखें- हर माह एक नया किस्सा होता है, जिसमें फेक आई डी और ब्लैकमेलिंग का जिक्र रहता है। कहीं कोई अपने नाम के दम पर किसी को छपवाने के दम्भ में डूबता नजर आता है तो कोई अपने गुट को रेवड़ी की तरह पुरस्कार बांटते हुए उन्हें अंतर्राष्ट्रीय साहित्यकार घोषित कर देता है। दुर्भाग्यपूर्ण है कि आप वर्षों से जिस क्षेत्र को सम्मान के साथ देखते आये हैं, उसकी कलई इससे जुड़ जाने के बाद ही खुलती है। लोग आपको उस हद तक गिराने के लिए तैयार हैं, जिस हद तक आप गिर सकते हैं। दोषी दोनों ही पक्ष हैं।

स्वाभिमानी बनें एवं पुरस्कार के लिए 'बिकाऊ' न हो जाएँ- दुनिया चाहे जो भी कहे, पर एक सच ये भी है कि अपनी क्षमता और लेखनी के स्तर का अंदाजा हर रचनाकार को होता है। 'तुम मुझे कविता दो, मैं तुम्हें पुरस्कार दूंगा', की तर्ज पर पत्रिका के हर अंक पर बँटते पुरस्कारों के क्या मायने हैं, ये पाने वाले को भी पता होते हैं। क्या 'साहित्य' की गिरती साख को बचाने और इस प्रदूषित वातावरण को थोड़ा स्वच्छ बनाने के लिए अब इन लोगों से मुक्ति का समय नहीं आ गया है?

अच्छे पाठक भी बनें- सूरज अपने प्रकाश का बखान नहीं करता और न ही चाँदनी अपनी सुंदरता पर मुग्ध हो इतराती है। फूलों की सुगंध से वातावरण अपने-आप महकता है और ठंडी हवा की शीतलता स्वयं ही चित्त में ताजगी भर देती है। प्रत्यक्ष को प्रमाण क्या! लेकिन फिर भी कई साहित्यकार ऐसे हैं जो कि आत्ममुग्धता में लिप्त हो, अन्य लेखकों को है, दृष्टि से देखते हैं। ये उन्हें अपनी पुस्तकें बेचकर उस पर समीक्षा लिखने की अपेक्षा तो बहुत करते हैं पर किसी नवोदित की पुस्तक में इन्हें कोई रुचि नहीं होती!

समाज में रहते हुए साहित्यकार, सम-सामयिक घटनाओं एवं परिस्थितियों से इतर होकर नहीं लिख सकता। आने वाली पीढ़ियों के लिए उसके शब्द उस कालखंड का प्रतिनिधि बन उभरते हैं। अतः यह उसका नैतिक दायित्व है कि वह अपने समय की समुचित समीक्षा कर उसे लेखनीबद्ध करता रहे। परन्तु वर्तमान परिवेश में यह भी

आवश्यक है कि वह समस्त घटनाओं पर पैनी दृष्टि रखते हुए प्रकृति संरक्षण, पर्यावरण, जीवन-संगीत पर भी लिखे। उसे देश की एकता, अखंडता, संस्कृति एवं तमाम सामाजिक मूल्यों के भाव संरक्षित रखने के साथ ही, उन विषयों पर भी कहना होगा जिनसे इस सृष्टि की सुंदरता रची गई है। उसे उन सभी सकारात्मक कहानी-किस्सों को लिखते रहना होगा जिससे समाज में आशा का भाव प्रबल हो और यह विश्वास भी संचित रहे कि दुनिया की अथाह सुन्दरता अब भी बनी हुई है और सभ्यता का दौर निरंतर जारी है।

हर प्रेम को सिद्ध नहीं करना होता है!

इन दिनों आप जिस भी दिशा में दृष्टिपात करेंगे, आपकी आँखों में निराशा की झलक ही लौटकर आएगी। हृदय उस उदासी का साक्षी बनेगा जो आशाओं के तिलिस्म से बाहर आ और भी गहराती चली जाती है। समाज की बदलती तस्वीर पर अब गर्व नहीं होता बल्कि मस्तक शर्मिंदगी से झुक जाता है। यह वही देश है जो विश्वगुरु की राह पर चलने का दावा करता है। यह वही देश है जो कभी सोने की चिड़िया कहा जाता था। यह वही देश है जिसकी सभ्यता और संस्कृति की दुहाई विश्वभर में सदियों से दी जाती रही है। तमाम विविधताओं के मध्य भी यहाँ की एकता, अखंडता और धर्मनिरपेक्षता का उजला चेहरा सदैव ही चमचमाता रहा है। दुःख और क्षोभ का विषय है कि हमारे अपने इसी देश में प्रेम के सुर्ख महीने में प्रेम ही लापता है। जहाँ वातावरण गुलाब की सुगंध से मन-मस्तिष्क को सुवासित कर देता है, शांति और अहिंसा की प्रतिमूर्ति बने उसी देश में, उसी की कनपटी पर बंदूक रख हिंसा का सामान्यीकरण होता जा रहा है।

इन मुट्ठी भर लोगों को यह समझना होगा कि हम देशवासियों में वो प्रेम अब भी जीवित है जो सबको जोड़ता आया है। हम उस संस्कृति के पूजक हैं जिसने हमें अपने सभी त्योहारों से बाँधे रखा है। होली, दीपावली, ईद, क्रिसमस, लोहड़ी, नवरोज, गुरुपर्व या कोई भी त्योहार होय हम एक सच्चे भारतीय की तरह उसे सबके साथ मनाते हैं।

उत्सवों ने हमें केवल उमंग और उल्लास भरे पल ही नहीं दिए बल्कि प्रेम के उस वृहद रूप से भी मिलवाया है जो मनुष्यता जीवित रखता है, मानव को मानव होने का मूलमंत्र देता है, भाईचारा बनाये रखता है।

हमारी भारतीयता उन दिनों की ऋणी है जिसने हममें मानवता के बीज बोए, जिसने सिखाया कि हम सब एक ही हैं, हाड़माँस के बने वो इंसान जिनके लहू का रंग भिन्न वस्त्र पहन लेने या भोजन बदल जाने से रत्ती भर भी परिवर्तित नहीं होता। हमारी मुस्कानें एक हैं, हमारे दुःख एक हैं, हमारी शिकायतें और कर्तव्य भी एक ही हैं। हमें उन चंद लोगों की बातों को अस्वीकृत करना होगा जो हमारी भारतीयता का हिन्दू-मुस्लिमीकरण करने में लगे हैं। यह देश हम सबका है और ये बात हमारे जन्म के समय से सिद्ध है। इसे तय करने का हक किसी और का कभी नहीं हो सकता! इस देश की माटी में हमारे बचपन की साँसें घुली हुई हैं। हमारा जीवन इस देश की आबोहवा का गुलाम है। हमने इस धरती पर उम्मीदों के

बीज बोये हैं, इसके आसमान पर हमारी चाहतों का चाँद उगाया है। यहाँ के पक्षियों से हमने अपने स्वप्न साझा किये हैं। पहाड़ों पर जा इस सुन्दर धरा को सौ-सौ बार नमन किया, इसे अपना कह पुकारा। हम इस ज़मीन को चूमकर दिन का प्रारम्भ करते हैं। यहाँ की नदियों में हमारी सरसराती उमंगों की नाव बहती है। हमारी प्रत्येक वाँस जिसके जंगलों, पठारों, प्राकृतिक छटा के प्रेम में रची-बसी है, उसे सिद्ध करने के लिए किसी कागज की नहीं बल्कि महसूसने के लिए मात्र एक निश्छल हृदय की दरकार है। यही हम भारतीयों का प्रेम है, यही हमारा देशप्रेम है जो स्वार्थ और थोथे दिखावे के हर भाव से हजारों गुना ऊपर है।

हम प्रकृति के कण-कण से प्रेम करने वाले पूर्वजों की संतान हैं। हम ये भी जानते समझते हैं कि हम किसी भी घर में जन्म ले सकते थे, कोई और भी हो सकते थे इसलिए हमने किसी का धर्म-जाति न देख केवल मनुष्यता को चुना। हमारे प्रेम की यही परिभाषा है जो दुर्भाग्य से उन नेताओं की परिभाषा से मेल नहीं खाती जो राष्ट्रप्रेम की बातें तो खूब करते हैं पर उसे समझते नहीं या यूँ कहें कि वे चाहते ही नहीं कि हम सब इसे समझ सकें। हमें एक क्षण को भी नहीं भूलना चाहिए कि जो लड़ाने की बात करे, दो इंसानों के बीच खाई पैदा करे, आपको भारतीय न कहकर हिन्दू-मुस्लिम करे वह मानवता का कट्टर दुश्मन है।

हमें खतरा केवल आततायियों से ही नहीं है बल्कि बढ़ते अपराधों के प्रति हमारी सहजता भी इस प्रवृत्ति को बढ़ावा दे रही है। देश की ही संपत्ति को नष्ट करके, निर्दोषों पर हमले कर 'देशभक्ति' की नई परिभाषाएँ गढ़ी जा रही हैं। कैमरे पर कैद अपराधियों को कभी कोई भय रहा ही नहीं वे जानते हैं उनके आका उन्हें बचा ही लेंगे। ऐसे लोग स्वहित के लिए जी रहे हैं इन्हें न आपसे प्रेम है, न देश से। यथाशीघ्र इनको प्रेम सहित नमस्कार कर दूरी बना लें। हमारे लिए किसी भी व्यक्ति विशेष से पहले हमारा देश है।

हमारी भाषा, प्रेम की ही भाषा है जिसे हमारे दिल से कोई जुदा नहीं कर सकता! और हाँ, हर प्रेम को सिद्ध नहीं करना होता है!

मेरी उम्मीदों की वसीयत!

अनिश्चितता से भरे जीवन में दुख क्यों बांटू? वसीयत इसलिए लिख रही हूं, क्योंकि कोरोना की दूसरी लहर का कहर हम सबने जिया है। ये कोई डरने या डराने का प्रयोजन नहीं है। हालात सबके सामने हैं। कोई दिन ऐसा नहीं जाता जब किसी के निधन का समाचार न मिलता हो। शायद ही कोई ऐसा घर होगा, जहाँ कोविड ने किसी को चपेट में न लिया हो। मेरे आसपास, परिवार, मित्र सभी की यही कहानी है। किसी का कोई अपना जूझ रहा है तो कहीं, कोई अलविदा कहे बिना चला गया। हर दिन एक आशंका में बीतता है कि वो जो अपनी साँसों के लिए लड़ रहा है, वो घर तो आ जाएगा न! उसे समय पर सही उपचार मिल जाएगा न! कहीं आज भी तंत्र की बलि तो न चढ़ जाएगा कोई! स्वयं को इतना लाचार और विवश पहले कभी भी न महसूस किया था।

ये भय इतना गहरा है कि सोशल मीडिया पर भी यूं लगने लगा है जैसे किसी अंतहीन शोक सभा में विराजमान हूँ। किसी की तस्वीर दिखते ही लपककर 'बहुत सुंदर' या 'नाइस पिक' नहीं लिखती हूँ बल्कि धडकनें इस हद तक बढ़ जाती हैं कि क्या कहूँ! अत्यधिक घबराहट से भर, धड़कते हृदय से पहले पोस्ट पढ़ती हूँ, कि कहीं...! कोई नियमित पोस्ट डालने वाला व्यक्ति अचानक गायब हो जाए, तब भी मन संशय से भर उठता है। कुछ को तो मैंने संदेश भी भेजे हैं। जिनके प्रत्युत्तर नहीं मिले हैं उन्हें लेकर बहुत चिंतित हूँ। अब फोन पर बातों में 'और सब ठीक ठाक है?' पूछने का एक ही प्रयोजन होता है कि 'किसी को कोरोना तो नहीं हुआ है न!' एक अजीब सी बेचौनी मस्तिष्क को घेरे रहती है। होंठ मंत्रोच्चार कर तमाम प्रार्थनाएं बुदबुदाने लगे हैं।

देखा जाए तो इन दिनों अधिकांश देशवासियों की तरह, जीवित बने रहने के सारे उपक्रम, मेरी भी दिनचर्या में जुड़ चुके हैं। साँसों की गति न थमे, इसके लिए योग किया जा रहा, साथ ही प्रतिरोधक शक्ति बढ़ाने के समस्त उपाय भी अपना रही हूँ। फिर चाहे वो काढ़ा हो या मल्टी विटामिन। फलों एवं वनस्पतियों का सेवन भी बढ़ा दिया है। कुल मिलाकर इस अदृश्य विषाणु से स्वयं को छुपाने और बचाने के जितने तरीके हैं, वो सब अपना रही हूँ। लेकिन साथ ही डर भी है कि न जाने कब, किस दिशा से वो आ धर दबोचे, तो सुरक्षा हेतु मैंने तो इस लड़ाई में अपनी पूरी शक्ति झोंक दी है। बहुत से लोग कहते हैं कि 'अरे, जीवन से इतना मोह किसलिए! 'होइहि सोइ जो राम रचि राखा'। मैं इस बात पर आंशिक सहमति

प्रदर्शित करते हुए इतना ही कहती हूँ कि अभी बहुत कुछ शेष है जो जिया जाना है, कुछ स्वप्न भी हैं जिन्हें पूरा करना है। जीवन को मैंने सदैव ही अमूल्य माना है और इसे ऐसे ही अपनी किसी गलती के चलते नहीं हारना चाहती। इसलिए वैक्सीन भी लगवाई है और सुरक्षा को लेकर पूर्णतः सतर्क भी रहती हूँ।

सचमुच यह अत्यंत विकट स्थिति है कि एक अति-सूक्ष्म जीव के आगे पूरा विश्व नतमस्तक है। अर्थव्यवस्था तहस-नहस हो चुकी, मानव जीवन नष्ट हो रहे, सामाजिक व्यवस्थाएं छिन्न-भिन्न हैं और हम अपने ही घर के किसी कोने में दुबककर रहने को विवश है। सभी प्राथमिकताओं को ताक पर रखकर, जीवन सुरक्षित रखना ही एकमात्र ध्येय बन चुका है। शेष रहेंगे, तब ही तो स्वप्न साकार कर सकेंगे।

इस समय एक भविष्यवाणी याद आ रही है जिसे मैंने (संभवतः आपने भी) बचपन से लेकर अब तक अनगिनत बार सुना है कि फलाने वर्ष में प्रलय आएगा। जनता त्राहिमाम कर उठेगी। हर तरफ लाशों के ढेर होंगे और धरती पर जीवन समाप्त हो जाएगा। मैं सदैव इसे हास्य की दृष्टि से ही देखती रही हूँ। आपकी भी स्मृति में होगा कि कैसे सब हँसा करते थे कि 'अरे! प्रलय आया नहीं अभी तक! उसकी तारीख तो निकल गई!'

हमने कल्पना भी न की थी कि सचमुच एक समय ऐसा आएगा, जब नित शोक समाचार ही प्राप्त होते रहेंगे, आँखें अनवरत भीगी रहेंगी और हाथ श्रद्धांजलि के लिए जुड़े ही रह जाएंगे। मैं मानने लगी हूँ कि कोविड महामारी ही वो प्रलय है जिसके बारे में हमें वर्षों से सचेत किया जाता रहा है।

लोग कहते हैं कि इस कठिन समय में नकारात्मक बातें नहीं होनी चाहिए। किसको अच्छी लगती हैं? लेकिन चिताओं के बीच में बैठ चेहरे पर मुस्कान बनाए रखना इतना सरल नहीं होता। दुखी मन से आशा की किरणें नहीं फूट पातीं। वह इंसान जिसने अपनों को खोया है, इलाज के लिए दर-दर भटकते, दम तोड़ते लोग देखे हैं। आखिर कब तक सकारात्मक बना रहे? जो सच है, उसे स्वीकार कर लेने में क्या आपत्ति है? डर इसलिए भी आवश्यक है कि हम निश्चिन्त होकर न बैठ जाएँ एवं अतिरिक्त सावधानी बरतें।

अब तो फेसबुक ने भी पूछना शुरू कर दिया कि आपके जाने के बाद आपके अकाउंट का क्या करें! पता नहीं, फेसबुक में ये सेटिंग कब से है लेकिन मेरी नजर इस पर हाल ही में पड़ी। इसमें यह विकल्प भी है कि मैं पहले से ही बता दूँ कि मेरे गुजर जाने के बाद मेरे अकाउंट को कौन देखेगा और वह श्रद्धांजलि की पोस्ट को कैसे मैनेज करेगा! यदि मैं अपनी मृत्यु के बाद अकाउंट

समाप्त करवाना चाहती हूँ तो वह सेवा भी उपलब्ध है। उसे देखती हूँ तो लगता है कि जैसे सोशल मीडिया खाते का 'जीवन बीमा' करा रही हूँ।

तो अब काम की बात–

जीवन की क्षणभंगुरता से ऐसे कभी भी साक्षात्कार न हुआ जो कोरोना ने कराया। इसलिए जो सबसे जरूरी है वो यह कि अपनी उम्मीदों की वसीयत कर ली जाए। यदि सचमुच सकारात्मकता लानी है, तो सर्वप्रथम तो राजनीति से दूर रहा जाए क्योंकि सर्वाधिक जहर इसी का बोया हुआ है। यदि जाने–अनजाने में किसी का दिल दुखाया हो तो यही समय है कि उससे क्षमा मांग ली जाए। मन में किसी के प्रति दुर्भावना हो तो उसे सद्भावना में बदल दें। झगड़े के कारण किसी से बोलचाल बंद है तो उसकी ओर मित्रता का हाथ बढ़ा लिया जाए। किसी को उसकी गलती के लिए हृदय से क्षमा कर दिया जाए। और जिससे प्रेम हो, उसे जता दिया जाए। न तो हमारे मन पर कोई बोझ रहे और न किसी के मन में हम कोई कसक छोड़कर जाएं। कौन जाने, कल हममें से कोई हो न हो!

आप भी प्रसन्न मन से कीजिए न, एक ऐसी ही वसीयत। निराशा के घने बादल छँटेंगे, जीवन पर भरोसा और हौसला, दोनों लौट आएंगें। हृदय में बस यही एक भाव स्थायी रूप से बस जाएगा– 'सर्वे भवन्तु सुखिनः'।

प्रीति 'अज्ञात'
(मूल नामः प्रीति जैन)
लेखिका, कवयित्री, संपादक, ब्लॉगर, सामाजिक कार्यकर्ता

शिक्षाः एम. एससी. (वनस्पति शास्त्र), डिप्लोमा इन संस्कृत टीचिंग
जन्म स्थानः भिंड (म.प्र.)
वर्तमान निवास-स्थानः अहमदाबाद (गुजरात)

संप्रतिः
मार्च 2015 से वर्तमान- संस्थापक एवं संपादक, 'हस्ताक्षर' (हिन्दी साहित्य की मासिक वेब पत्रिका)
जुलाई 2016 से वर्तमान- इंडिया टुडे की वेबसाइट iChowk पर लेखन (contributor)
अक्टूबर 2019 से वर्तमान- साहित्य, कला एवं संस्कृति को समर्पित संस्था 'कर्मभूमि-अहमदाबाद' की सह-संस्थापक
जनवरी 2019 से वर्तमान- दिल्ली से प्रकाशित मासिक पत्रिका 'साहित्यनामा' में नियमित स्तम्भकार (प्रीत के बोल)
2013-14: संपादक (साहित्य रागिनी वेब पत्रिका)
2012 से वर्तमान/सक्रिय ब्लॉगर
लेखन यात्राः स्कूली दिनों से प्रारम्भ (1988)

प्रमुख विधाएं: कविता, कहानी, निबंध, व्यंग्य, यात्रा वृत्तांत, संस्मरण, सामाजिक-समसामयिक विषयक आलेख

प्रकाशित पुस्तकेंः
2022: व्यंग्य संग्रह - देश मेरा रंगरेज़ है, बाबू
2019: आलेख संग्रह- दोपहर की धूप में
2016: काव्य-संग्रह- मध्यांतर
2013-2020: पाँच पुस्तकों का सम्पादन, 15 साझा काव्य-संग्रह,

2 साझा संस्मरण-संग्रह, 1 कहानी-संग्रह में रचनाएँ प्रकाशित, एक दर्जन से भी अधिक पुस्तकों की भूमिका एवं समीक्षा लेखन
2012 से सक्रिय ब्लॉग लेखन- 2 ब्लॉग
1) ख़्वाहिशों के बादलों की कुछ अनकही, कुछ अनसुनी
2) यूँ होता...तो क्या होता
1988-1998: साप्ताहिक हिंदुस्तान, धर्मयुग के 'युवा जगत' स्तंभों में प्रकाशित, विभिन्न राष्ट्रीय पत्र-पत्रिकाओं में स्त्री विषयक परिचर्चाओं में शामिल, प्रादेशिक समाचार-पत्रों में रचनाएँ प्रकाशित।

उपलब्धियाँ/पुरस्कार:

'दोपहर की धूप में' पुस्तक गुजरात साहित्य अकादमी द्वारा श्रेष्ठ पुस्तक पारितोषिक (मौलिक गद्य लेखन श्रेणी में) के लिए चयनित (2021)

2020: लाडली मीडिया एंड एडवरटाइजिंग अवार्ड्स फॉर जेंडर सेंसिटिविटी 2020

2019: वर्ष के टॉप 10 ब्लॉगर ऑफ द ईयर में प्रथम स्थान प्राप्त

2018: सितम्बर 2018 में जयपुर में WOTFA पुरस्कार से सम्मानित

2018: पुस्तक मध्यांतर शांति-गया पुरस्कार से सम्मानित (सरोज खरे स्मृति पुरस्कार), जून 2018 भोपाल

2017: बड़ौदा में आयोजित यंग गुजरात कार्यक्रम में पुस्तक मध्यांतर के लिए Author Recognition Certificate से सम्मानित (मार्च 2017, बड़ौदा)

2017: स्पंदन महिला साहित्यिक एवं शैक्षणिक संस्थान, जयपुर द्वारा अखिल भारतीय डॉ. कुमुद टिक्कू कविता प्रतियोगिता में 'श्रेष्ठ कविता' पुरस्कार (जनवरी 2017)

2017: ग्रामीण पत्रकारिता विकास संस्थान, ग्वालियर (म.प्र.) द्वारा हिन्दी पत्रकारिता एवं साहित्यिक गौरव सम्मान (मई, 2017, ग्वालियर)

2015: सर्वश्रेष्ठ ब्लॉगर के लिए परिकल्पना साहित्य सम्मान 2015 (अंतर्राष्ट्रीय ब्लॉगर सम्मेलन, बैंकॉक, थाईलैंड)

1988: म.प्र पत्र-लेखक संघ द्वारा पुरस्कृत (1988)

साहित्यिक एवं सामाजिक कार्यों हेतु कई अन्य संस्थाओं द्वारा पुरस्कृत

विशेष:

जनवरी 2020 में दूरदर्शन पर साक्षात्कार प्रसारित

आकाशवाणी से कविताएं एवं FM रेडियो पर वार्ता प्रसारित

फरवरी 2020 में मुंबई विश्वविद्यालय द्वारा आयोजित बहुभाषी काव्य-गोष्ठी में कविता-पाठ

मार्च 2019 में गुजरात विद्यापीठ, अहमदाबाद में काव्य पाठ (जूही मेला-2019)

षष्टम अंतर्राष्ट्रीय ब्लॉगर सम्मेलन, बैंकॉक (थाईलैंड) में 'साहित्य की समृद्धि में महिलाओं का योगदान' विषय पर आलेख प्रस्तुति (जनवरी' 2016)

स्वामी केशवानंद राजस्थान कृषि वि.वि, बीकानेर (स्नातकोत्तर अध्ययन विभाग) द्वारा आयोजित राष्ट्रीय संगोष्ठी में 'महिला सशक्तिकरण चुनौतियाँ एवं संभावनाएँ' विषय पर आलेख प्रस्तुति (सितम्बर 2017)

प्रतिष्ठित वेबसाइट 'कविताकोश' में कविताएँ सम्मिलित

तृतीय एवं चतुर्थ 'दिल्ली अंतर्राष्ट्रीय फिल्म महोत्सव' के लिए कवितायें चयनित एवं प्रकाशित

विश्व पुस्तक मेला दिल्ली में, 2016,17 में 'गूँज' एवं 2016 में 'रेवान्त' के मंच से कविता-पाठ

संपर्क सूत्रः
मोबाइलः 9727069342
ई.मेलः preetiagyaat@gmail.com
वेबसाइटः https://hastaksher.com/